CHARACTERIZATION METHODS FOR SUBMICRON MOSFETs

THE KLUWER INTERNATIONAL SERIES IN ENGINEERING AND COMPUTER SCIENCE

ANALOG CIRCUITS AND SIGNAL PROCESSING
Consulting Editor
Mohammed Ismail
Ohio State University

Related Titles:

LOW-VOLTAGE LOW-POWER ANALOG INTEGRATED CIRCUITS, edited by Wouter Serdijn
ISBN: 0-7923-9608-1
INTEGRATED VIDEO-FREQUENCY CONTINUOUS-TIME FILTERS: *High-Performance Realizations in BiCMOS*, Scott D. Willingham, Ken Martin
ISBN: 0-7923-9595-6
FEED-FORWARD NEURAL NETWORKS: *Vector Decomposition Analysis, Modelling and Analog Implementation*, Anne-Johan Annema
ISBN: 0-7923-9567-0
FREQUENCY COMPENSATION TECHNIQUES LOW-POWER OPERATIONAL AMPLIFIERS, Ruud Easchauzier, Johan Huijsing
ISBN: 0-7923-9565-4
ANALOG SIGNAL GENERATION FOR BIST OF MIXED-SIGNAL INTEGRATED CIRCUITS, Gordon W. Roberts, Albert K. Lu
ISBN: 0-7923-9564-6
INTEGRATED FIBER-OPTIC RECEIVERS, Aaron Buchwald, Kenneth W. Martin
ISBN: 0-7923-9549-2
MODELING WITH AN ANALOG HARDWARE DESCRIPTION LANGUAGE,
H. Alan Mantooth, Mike Fiegenbaum
ISBN: 0-7923-9516-6
LOW-VOLTAGE CMOS OPERATIONAL AMPLIFIERS: *Theory, Design and Implementation*, Satoshi Sakurai, Mohammed Ismail
ISBN: 0-7923-9507-7
ANALYSIS AND SYNTHESIS OF MOS TRANSLINEAR CIRCUITS, Remco J. Wiegerink
ISBN: 0-7923-9390-2
COMPUTER-AIDED DESIGN OF ANALOG CIRCUITS AND SYSTEMS, L. Richard Carley, Ronald S. Gyurcsik
ISBN: 0-7923-9351-1
HIGH-PERFORMANCE CMOS CONTINUOUS-TIME FILTERS, José Silva-Martínez, Michiel Steyaert, Willy Sansen
ISBN: 0-7923-9339-2
SYMBOLIC ANALYSIS OF ANALOG CIRCUITS: Techniques and Applications, Lawrence P. Huelsman, Georges G. E. Gielen
ISBN: 0-7923-9324-4
DESIGN OF LOW-VOLTAGE BIPOLAR OPERATIONAL AMPLIFIERS, M. Jeroen Fonderie, Johan H. Huijsing
ISBN: 0-7923-9317-1
STATISTICAL MODELING FOR COMPUTER-AIDED DESIGN OF MOS VLSI CIRCUITS, Christopher Michael, Mohammed Ismail
ISBN: 0-7923-9299-X
SELECTIVE LINEAR-PHASE SWITCHED-CAPACITOR AND DIGITAL FILTERS, Hussein Baher
ISBN: 0-7923-9298-1
ANALOG CMOS FILTERS FOR VERY HIGH FREQUENCIES, Bram Nauta
ISBN: 0-7923-9272-8
ANALOG VLSI NEURAL NETWORKS, Yoshiyasu Takefuji
ISBN: 0-7923-9273-6
ANALOG VLSI IMPLEMENTATION OF NEURAL NETWORKS, Carver A. Mead, Mohammed Ismail
ISBN: 0-7923-9049-7
AN INTRODUCTION TO ANALOG VLSI DESIGN AUTOMATION, Mohammed Ismail, José Franca
ISBN: 0-7923-9071-7

CHARACTERIZATION METHODS FOR SUBMICRON MOSFETs

edited by

Hisham Haddara
Ain Shams University

KLUWER ACADEMIC PUBLISHERS
Boston / Dordrecht / London

Distributors for North America:
Kluwer Academic Publishers
101 Philip Drive
Assinippi Park
Norwell, Massachusetts 02061 USA

Distributors for all other countries:
Kluwer Academic Publishers Group
Distribution Centre
Post Office Box 322
3300 AH Dordrecht, THE NETHERLANDS

Library of Congress Cataloging-in-Publication Data

A C.I.P. Catalogue record for this book is available
from the Library of Congress.

Contents

List of Contributors

Gérard Ghibaudo	LPCS, UA-CNRS & INPG, ENSERG, Grenoble, France
Hisham Haddara	Ain Shams University, Cairo, Egypt
C.R. Viswanathan	University of California at Los Angeles, California, USA
Peter McLarty	North Carolina State University, North Carolina, USA
H. H. Mueller	University of Erlangen, Erlangen, Germany
Max Schulz,	University of Erlangen, Erlangen, Germany
Sorin Cristoloveanu	LPCS, UA-CNRS & INPG, ENSERG, Grenoble, France
Hing-Yan To	Ohio State University, Ohio, USA
Mohamed Ismail	Ohio State University, Ohio, USA

Preface

It is true that the Metal-Oxide-Semiconductor Field-Eeffect Transistor (MOSFET) is a key component in modern microelectronics. It is also true that there is a lack of comprehensive books on MOSFET characterization in general. However there is more than that as to the motivation and reasons behind writing this book. During the last decade, device physicists, researchers and engineers have been continuously faced with new elements which made the task of MOSFET characterization more and more crucial as well as difficult. The progressive miniaturization of devices has caused several phenomena to emerge and modify the performance of scaled-down MOSFETs. Localized degradation induced by hot carrier injection and Random Telegraph Signal (RTS) noise generated by individual traps are examples of these phenomena. Therefore, it was inevitable to develop new models and new characterization methods or at least adapt the existing ones to cope with the special nature of these new phenomena.

The need for more deep and extensive characterization of MOSFET parameters has further increased as the applications of this device have gained ground in many new fields in which its performance has become more and more sensitive to the properties of its $Si - SiO_2$ interface. MOS transistors have crossed the borders of high speed electronics where they operate at GHz frequencies. Moreover, MOSFETs are now widely employed in the subthreshold regime in neural circuits and biomedical applications. In both of these two domains - high speed and low current (subthreshold regime) - the device performance is strongly influenced by the quality of its interface.

Last but not least, the appearance of new materials especially Silicon-On-Insulator (SOI), and the subsequent development of new device structures have strongly called for new characterization methods and procedures.

This book has been written with the intention of providing some help to device engineers and researchers in order to enable them to cope with such challenges. Without adequate device characterization, new physical phenomena and new types of defects or damage may not be well identified or dealt with, leading to an undoubted obstruction of the device development cycle.

The primary audience of this book falls into two categories: graduate students who are familiar with MOS device physics and wish to work in the field of device characterization and modeling. The second group is industrial engineers who are working in device development and want to enlarge their scope of knowledge of measurement methods. Moreover, The book also adresses device-based characterization for material and process engineers as well as for circuit designers.

The contributors to this book have first hand practical experience with the subjects they have treated and are known experts in their fields. The approach they adopted was to give proportionate weights to the theoretical foundations and practical aspects of different characterization techniques.

This book does not claim to be an exhaustive survey of existing MOSFET characterization methods. Instead, it deals only with those techniques which show high potential for characterization of submicron devices. Focus is made throughout the book on the adaptation of such methods to resolve measurement problems relevant to VLSI devices and new materials especially SOI.

The first chapter deals with modeling of the static characteristics of VLSI MOSFETs and parameter extraction. It also serves as an introduction for this book where the different phenomena affecting the modeling and operation of sub-μm MOSFETs are addressed. The chapter begins by highlighting the motivation and need for extraction of MOSFET parameters (e.g. threshold voltage, channel mobility, series resistance etc...). This is followed by an overview of dc models in different bias regimes. The rest of the chapter is devoted to measuring techniques and parameter extraction methods at room and low temperatures.

The second chapter discusses small signal modeling and characterization of MOSFETs. Variants of different ac models of the MOS transistor are presented and analyzed. The frequency response of MOSFETs is discussed and the parameters which govern the speed of operation of these devices are pointed out. Models of the split-admittance and dynamic transconductance are then derived and methods of extracting interface trap parameters are presented. Also discussed at the end of this chapter, is the application of small signal measurements for the determination of other MOSFET parameters.

Chapter three is devoted to charge pumping (CP) ; one of the most reliable methods for interface characterization in MOSFETs. The charge pumping phenomenon and early experiments on charge pumping are first reviewed. Next, interface trap kinetics during the charge pumping cycle are analyzed. The application of the CP technique for interface trap characterization is then illustrated and different variants of the technique are discussed. In the remaining part of the chapter, special aspects and applications of charge pumping are emphsized in detail ; this involves charge pumping in submicron devices, charge pumping at low tempratures and characterization of stressed and damaged devices using the CP technique.

The subject of the fourth chapter is Deep Level Transient Spectroscopy (DLTS) in MOS transistors and in particular current-DLTS (a technique which has lately found significant interest in the area of SOI devices). The chapter begins with a discussion on generation, recombination and trapping statistics and models. Theoretical basis of MOSFET current transient spectroscopy is then presented followed by models and methods of bulk trap and interface trap characterization. Details of the instrumentation and the different current transient measurement techniques are provided at the end of this chapter.

Chapter five deals with individual interface traps and telegraph noise in submicron MOS transistors. It starts with an overview on the origin of random telegraph signals and a discussion of emission and capture times, emission transients, noise and complex random telegraph signals. The experimental properties of individual traps are then identified and the remaining part of the chapter is spared for the analysis, modeling and evaluation of measurements.

The last two chapters are not assigned to specific characterization methods. However, they have been reserved for two important issues of wider scope. The characterization of VLSI MOSFETs fabricated on SOI material forms the subject of chapter six whereas the last chapter addresses device-based characterization of the CMOS process for analog circuit design.

Chapter six first introduces the reader to the world of SOI devices explaining the interest and applications of this technology. The fabrication, synthesis and mechanisms of operation of SOI structures are then briefly presented. After that, different characterization techniques are analyzed from the point of view of their applicability to SOI devices. The chapter ends with a discussion on hot carrier induced degradation in SOI MOSFETs.

Chapter seven has an objective of treating special concerns of analog circuit designers. Subjects like mismatch and mismatch drift, characterization of diffused and polysilicon resistors, precision capacitors and MOS transistors are discussed in detail. The implications for circuit and layout design are pointed out and analyzed. This chapter also includes a part dealing with statistical modeling and parameter extraction of BJTs ; these models are used side by side with those of MOS transistors in order to analyze BiCMOS circuits whose applications have rapidly been gaining grounds during the last few years.

I gratefully acknowledge the contributions of the authors of this book and I hope it will be of some help to its readers whether they are students, researchers or engineers working in the field of MOS devices and technology.

1

Static Measurements and Parameter Extraction

Gérard Ghibaudo
Laboratoire de Physique des Composants à Semiconducteurs
Grenoble, France

1.1 Introduction

The Metal-Oxide-Semiconductor field effect transistor (MOSFET) is one of the key devices for the fabrication of very (or ultra) large scale integrated circuits in modern microelectronics. The performance of the MOSFET is primarily determined by the quality of the gate dielectrics and that of the $Si - SiO_2$ interface which directly affects the carrier transport properties. On the other hand, the modeling of the device characteristics requires the rigorous definition of the MOSFET parameters which mainly control the device operation. Furthermore, the design of analog and digital circuits relies on electrical simulations based on SPICE-like programs in which state-of-the-art MOSFET static models have to be implemented in analytical forms. For this reason, the modeling of the submicron MOS transistor is a mandatory issue for the development of new CMOS circuits and semiconductor memories. Besides, the static measurements and the corresponding parameter extraction of MOSFETs have proved to be a powerful and simple characterization tool even though they are not competing with other electrical techniques also presented in this book.

The MOS transistor has been the subject of much research during the last decades. Many studies have contributed to increase our understanding of the physics and modeling of the MOSFET operation [1-12]. As a result, many models providing the output and transfer characteristics have been proposed for the description of the MOSFET operation in different regimes (weak inversion, strong inversion, saturation, ...) [1-12].

The aim of this chapter is first to present in a generic way the basic equations necessary for the modeling of the static MOSFET operation. To this end, the relevant charges, potentials and drain current equations of the MOS transistor are briefly established. Moreover, the principal MOSFET parameters which govern the ohmic and non linear operation regions are defined. The various analytical approximations used in the strong and weak inversion regions are also discussed. Finally, the refinements introduced for the modeling of submicronic phenomena such as short channel effects, Drain Induced Barrier Lowering or channel length shortening are also addressed. In a second step, the methods currently used for the extraction and characterization of the main MOS parameters are presented. The concept of charge and extrapolated threshold voltages are clarified. The relevant mobility coefficients such as the low field mobility and the mobility attenuation factor are explained. The notions of effective electrical channel length and channel width are also presented. Finally, the basic concepts and experimental means for measuring the static MOSFET characteristics are briefly described.

1.2 Modeling of MOSFET DC Characteristics

In this section, the basic equations used for the modeling of the static MOSFET characteristics are given for an enhancement mode n-channel device. Since the drain current is a function of the inversion charge and of the effective mobility, we present successively the relevant relations for these quantities. Finally, the models currently used for the linear and saturation regions will be presented.

1.2.1 Inversion Channel Charge Calculation

The gate induced charge in the semiconductor can be obtained after solving the Poisson equation throughout the $Si - SiO_2$ system. The semiconductor charge is a function of the surface potential ψ_s (*i.e.* band bending) and can be decomposed as a sum of the depletion charge Q_d and the inversion charge Q_i. For a p-type substrate, they are approximately given by the following equations [13,14],

$$Q_d = -\sqrt{2q\epsilon_{si}N_A(\psi_s - V_b)} \tag{1.1}$$

and

$$Q_i = -\frac{1}{2}\left[\sqrt{Q_d^2 + \frac{4q\epsilon_{si}n_i^2}{N_A}\exp\left(\frac{q\psi_s}{kT}\right)} - Q_d\right] \tag{1.2}$$

where q is the absolute electron charge, ϵ_{si} the silicon permittivity, n_i the intrinsic concentration, N_A the substrate doping, kT/q the thermal voltage and V_b the substrate bias. The gate charge conservation equation allows one to relate the charges to the gate voltage applied to the device with the substrate being grounded,

$$V_g = V_{fb} + \psi_s - \frac{1}{C_{ox}}(Q_i + Q_d + Q_{it}) \tag{1.3}$$

where $Q_{it} = -qD_{it}\psi_s$ is the interface trap charge referred to the flat-band condition, V_{fb} is the flat-band voltage, $V_{fb} = \Phi_{ms} - Q_{it0}/C_{ox}$ with Q_{it0} being the sum of flat-band interface trap charge and the fixed oxide charge, D_{it} the interface trap density $(eV^{-1}cm^{-2})$ and, Φ_{ms} the gate to semiconductor work function difference. For N^+ poly Silicon gate, $\Phi_{ms} = -(E_g/2q + \Phi_f)$ with $\Phi_f = kT/q \ln(N_A/n_i)$ being the bulk Fermi potential referred to the intrinsic energy level E_i. The gate voltage at which the surface potential equals twice the Fermi potential corresponds to the inversion threshold *i.e.* where the surface electron density n_s equals the hole majority carrier concentration, $n_s = N_A$. The inversion threshold voltage V_t separates the weak inversion or subthreshold region from the strong inversion region. Since at strong inversion, the surface potential almost saturates at $2\Phi_f$, the inversion threshold voltage can therefore be approximated as [13,14],

$$V_t = V_{fb} + 2\Phi_f - \frac{1}{C_{ox}}(Q_d + Q_{it}) \tag{1.4}$$

As a result, at strong inversion, the inversion charge varies almost linearly with gate voltage,

$$Q_i \simeq -C_{ox}(V_g - V_t) \tag{1.5}$$

For this reason, the threshold voltage is also called the "charge" threshold voltage. Below threshold (*i.e.* in weak inversion), the inversion charge varies exponentially with gate voltage and can be expressed as [4],

$$Q_i \simeq -\left(\frac{kT}{q}\right) C_d \exp\left[\frac{qA}{kT}(V_g - V_t)\right] \tag{1.6}$$

where C_d is the depletion capacitance ($C_d = -dQ_d/d\psi_s$) and $A = C_{ox}/(C_{ox} + C_d + C_{it})$ with C_{it} being the interface state capacitance ($C_{it} = qD_{it}$). Therefore, a good analytical formula for the inversion charge which covers all the inversion regions can be the following,

$$Q_i \simeq \frac{kTC_{ox}}{qA} \ln\left\{1 + \exp\left[\frac{qA}{kT}(V_g - V_t)\right] + \ln(1 - A)\right\} \tag{1.7}$$

such that it reduces to Equation 1.5 at strong inversion and to Equation 1.6 below threshold.

1.2.2 MOSFET DC Operation Models

Linear Operation Region

In the linear region of operation, the MOS transistor behaves as a gate-controlled quasi two dimensional resistor and the drain current I_d is, in a general way, given by [13,14] :

$$I_d = \mu_{eff} \left(\frac{W}{L} \right) |Q_i| V_d \tag{1.8}$$

where μ_{eff} is the corresponding effective mobility, V_d the drain voltage, W the channel width and L the channel length. The effective mobility of the carriers in the channel has been the object of much research [15-17]. It is generally found that at room temperature and for surface electric fields not exceeding 3-4 MV/cm, the effective mobility is well described by a function of the inversion and depletion charges or effective electric field E_{eff} of the form,

$$\frac{1}{\mu_{eff}} = \frac{1}{\mu_{00}} + \frac{E_{eff}}{E_c} = \frac{1}{\mu_0} + \frac{|Q_i|}{Q_c} \tag{1.9}$$

where μ_{00} is the zero-field mobility, μ_0 is the low-field mobility (at threshold where $Q_i \simeq 0$), Q_c is a critical inversion charge ($Q_c \simeq 10^{13}\ q/cm^2$) and E_c is a critical electric field ($E_c \simeq 10^4\ V/cm$). The effective electric field is a fraction of the surface electric field defined as [14-16],

$$E_{eff} = \frac{\eta Q_i + Q_d}{\epsilon_{si}} \tag{1.10}$$

where η is the inversion charge to depletion charge weighting factor. η is generally found to be equal to 1/2 for electrons and 1/3 for holes at room temperature. The transconductance of the MOSFET, $g_m = dI_d/dV_g$, can be obtained from Equations 1.8 and 1.9 as [10] :

$$g_m = \left(\frac{W}{L} \right) C_{ox} V_d \frac{C_i}{C_{ox} + C_d + C_i + C_{it}} \frac{\mu_{eff}^2}{\mu_0} \tag{1.11}$$

where C_i is the inversion charge capacitance. Note that Equation 1.11 provides in a continuous way the transconductance from the weak to the strong inversion regimes. The corresponding field effect mobility μ_{fe} is then obtained by normalizing Equation 1.11 as :

$$\mu_{fe} = \frac{L}{W C_{ox} V_d} g_m \tag{1.12}$$

It is worth emphasizing that the field effect mobility is not to be confused with the actual effective mobility. In strong inversion, $C_i \gg (C_{ox} + C_d + C_{it})$ so that $\mu_{fe} \simeq \mu_{eff}^2/\mu_0$. In weak inversion, $C_i \ll (C_{ox} + C_d + C_{it})$ and therefore μ_{fe} varies essentially as C_i (i.e. as the inversion charge since $C_i \simeq -qQ_i/kT$). Typical variations of the field effect and effective mobilities with

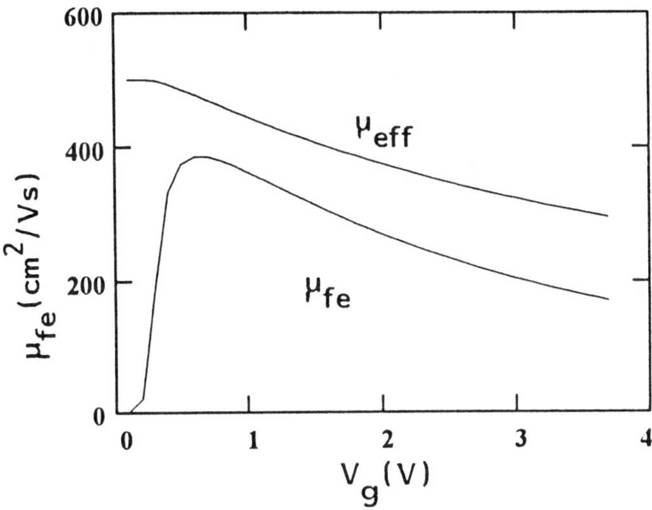

Figure 1.1: Typical variations of the field effect μ_{fe} and effective μ_{eff} mobilities with gate voltage V_g (parameters : $\mu_0 = 500\ cm^2/(Vs)$, $t_{ox} = 10\ nm$, $N_A = 10^{16}/cm^3$, $V_{fb} = -0.6\ V$).

gate voltage are presented in Figure 1.1. This comparison clearly points out that, in strong inversion, μ_{fe} is close to μ_{eff} but is always smaller (at least at room temperature) and that, in weak inversion, the magnitude of μ_{fe} is strongly different from that of the effective mobility.

In weak inversion, $\mu_{eff} \simeq \mu_0$ such that the drain current reads [4] :

$$I_d = \mu_0 \left(\frac{W}{L}\right) C_d \frac{kT}{q} \exp\left[\frac{qA(V_g - V_t)}{kT}\right] V_d \tag{1.13}$$

Therefore, the transconductance in weak inversion varies exponentially with gate voltage and is proportional to the current as $g_m = (qA/kT)I_d$. This allows one to evaluate experimentally the subthreshold slope from the ratio g_m/I_d. Combining Equation 1.5 with Equations 1.8 and 1.9 enables the drain current I_d to be expressed in strong inversion under the usual form [13,14],

$$I_d = \mu_0 \left(\frac{W}{L}\right) C_{ox} \frac{V_g - V_t}{1 + \theta(V_g - V_t)} V_d \tag{1.14}$$

where $\theta = C_{ox}/Q_c$ is the "mobility attenuation factor". The transconductance in strong inversion is thus given by

$$g_m = \left(\frac{W}{L}\right) C_{ox} \frac{\mu_0}{[1 + \theta(V_g - V_t)]^2} V_d \tag{1.15}$$

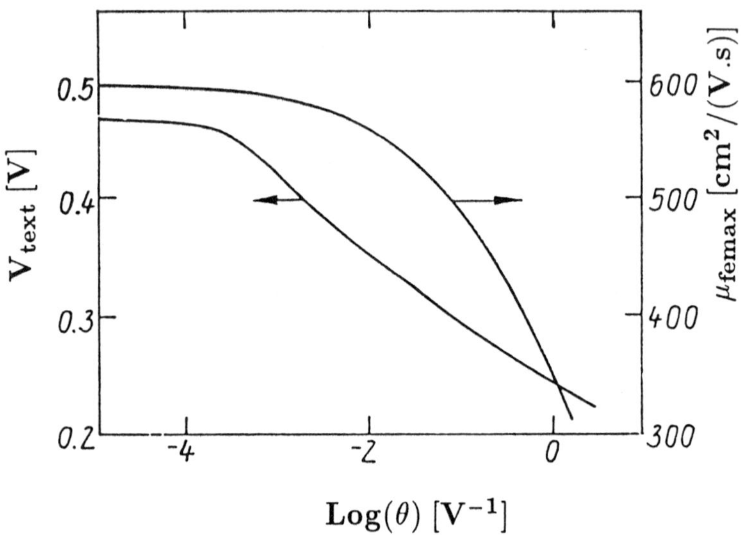

$$\mathbf{Log}(\theta)\ [\mathbf{V^{-1}}]$$

Figure 1.2: Variations of the extrapolated threshold voltage V_{text} and maximum field effect mobility μ_{femax} with mobility reduction factor θ (parameters : $\mu_0 = 600\ cm^2/(Vs)$, $t_{ox} = 26\ nm$, $N_A = 10^{16}\ /cm^3$).

At this point, it is interesting to mention that the charge threshold voltage V_t is not equivalent to the extrapolated threshold voltage V_{text} that can be directly deduced from the inflection point of the $I_d(V_g)$ characteristics. In fact, V_{text} is related to the inflection point coordinates by [10,12] :

$$V_{text} = V_{gmax} - \frac{I_d(V_{gmax})}{g_m(V_{gmax})} \tag{1.16}$$

where V_{gmax} is the gate voltage at which g_m is maximum. It can be shown using Equation 1.16 that V_{text} is linked to V_t, V_{gmax} and θ by the relationship [12] :

$$V_{text} = V_t - \theta(V_{gmax} - V_t)^2 \tag{1.17}$$

This results in the fact that V_{text} is always smaller than V_t as long as θ is different from zero (*i.e.* the effective mobility is not constant with V_g). Figure 1.2 illustrates the influence of θ both on the values of the extrapolated threshold voltage V_{text} and on the maximum field effect mobility μ_{femax}.

When the source-drain series resistance, R_{sd}, has to be taken into account, which is the case for short channel devices, one has to replace, in the preceding set of equations, the intrinsic mobility reduction factor θ by an effective one: $\theta^* = \theta + W\mu_0 C_{ox} R_{sd}/L$ [12]. So, an increase of the series resistance will result in an apparently accentuated mobility reduction and in a decrease of the extrapolated threshold voltage [12].

To some extent, short-channel effects can be analytically taken into account by evaluating the shift of the threshold voltage ΔV_t due to the surface potential increase in the middle of the channel because of the source and drain proximity. A lot of expressions for ΔV_t has been derived in the literature [6,18]. For example, in the trapezoidal charge sharing model, ΔV_t is given by [6] :

$$\Delta V_t = -\frac{qN_A W_d X_j}{C_{ox} L}\left[\left(1 + \frac{2y_s}{X_j}\right)^{1/2} - 1\right]$$

(1.18)

where W_d is the maximum depletion width of the source (drain)-substrate diode, X_j the source (drain) junction depth and y_s the depletion layer width at the surface. This relation holds for channel lengths down to $\approx 0.5 - 0.6\ \mu m$. Beyond this limit, more sophisticated formulae have been worked out which predict a stronger decreasing rate with channel length. For example, solving the pseudo two dimensional Poisson equation in the channel enables the threshold voltage shift to be expressed as [19],

$$\Delta V_t = -\frac{(V_{bi} - 2\Phi_f)}{\cosh(L/2l) - 1}$$

(1.19)

where $l = 2\sqrt{t_{ox} X_j/\xi}$ with ξ being a fitting parameter close to unity and V_{bi} the built-in potential between the source and substrate. Equation 1.19 is expected to be applicable for channel lengths down to 0.1-0.15 μm [19].

Non Linear Operation Regime

In the non linear regime of operation, the drain current can be obtained in a general way from the integration over the quasi-Fermi level shift Φ_c between source and drain as [12-14] :

$$I_d = \frac{W}{L}\int_0^{V_d} \mu_{eff}(Q_i)|Q_i(\Phi_c)|d\Phi_c$$

(1.20)

In weak inversion, I_d is then obtained from Equation 1.20 and from the charge conservation equation in the form [4] :

$$I_d = I_{dsat}\left[1 - \exp\left(-\frac{qCV_d}{kT}\right)\right]$$

(1.21)

where $I_{dsat} = W\mu_0 C_d kT^2/(q^2 CL)\exp[qA(V_g - V_t)/kT]$ is the saturation current; A is defined in Equation 1.6 and $C = (C_{ox} + C_d)/(C_{ox} + C_d + C_{it})$. The output conductance $g_d = dI_d/dV_d$ is then deduced from Equation 1.21 as :

$$g_d = G_d\exp\left(-\frac{qCV_d}{kT}\right)$$

(1.22)

with $G_d = g_d(0) = W\mu_0 C_d kT/(qL)\exp[qA(V_g - V_t)/kT]$ being the ohmic region conductance.

In strong inversion, $Q_i(\Phi_c) \simeq C_{ox}(V_g - V_t - \Phi_c)$ and the effective mobility is $\mu_{eff} = \mu_0/(1 + \theta(V_g - V_t - \Phi_c))$ so that the drain current can be obtained from Equation 1.20 as [12] :

$$I_d = \frac{W}{L}\mu_0 C_{ox}\frac{V_d}{\theta}\left[1 + \frac{1}{\theta V_d}\ln\left(\frac{1 + \theta(V_g - V_t - V_d)}{1 + \theta(V_g - V_t)}\right)\right] \qquad (1.23)$$

Equation 1.23 is valid below saturation (i.e for $V_d < V_{dsat}$; $V_{dsat} = V_g - V_t$). For $V_d \geq V_{dsat}$, I_d is constant and equal to the saturation current :

$$I_{dsat} = \frac{W}{L}\mu_0 C_{ox}\frac{V_g - V_t}{\theta}\left[1 - \frac{1}{\theta(V_g - V_t)}\ln\left(1 + \theta(V_g - V_t)\right)\right] \qquad (1.24)$$

Note that (1.23) and (1.24) have been derived taking into consideration the mobility gate voltage dependence which is not the case of the classical model where the mobility is assumed to be constant (see e.g. [13-14]). The corresponding saturation transconductance g_{msat} is therefore deduced from (1.24) as :

$$g_{msat} = \frac{W}{L}\mu_0 C_{ox}\frac{V_g - V_t}{1 + \theta(V_g - V_t)} \qquad (1.25)$$

Likewise, the output conductance in strong inversion and below saturation is found from (1.23) as :

$$g_d = \frac{W}{L}\mu_0 C_{ox}\frac{V_g - V_t - V_d}{1 + \theta(V_g - V_t - V_d)} \qquad (1.26)$$

In the case of short-channel devices, it is necessary to take into account the velocity saturation phenomenon along the channel. This results in a decrease of the mobility with drain voltage that can be approximated by [1,13,14] :

$$\mu_0 = \frac{\mu_{00}}{1 + KV_d} \qquad (1.27)$$

where $K = \mu_{00}/(Lv_{sat})$ is the longitudinal mobility reduction factor and v_{sat} is the saturation velocity. The main consequence of the saturation velocity effect is to decrease both the drain saturation voltage and the normalized drain saturation current. An analytical expression for V_{dsat} can be deduced from Equations 1.20 and 1.27 when the mobility reduction factor θ is neglected. Then, V_{dsat} is given by [2,12,14]:

$$V_{dsat} = \frac{-1 + \sqrt{1 + 2K(V_g - V_t)}}{K} \qquad (1.28)$$

Figure 1.3 shows the typical dependence of V_{dsat} on gate voltage as obtained using Equation 1.28 for different channel lengths. So, for relatively long channel devices, V_{dsat} is nearly equal to $(V_g - V_t)$ whereas, for shorter ones, it varies sublinearly with $(V_g - V_t)$. In fact, Equation 1.28 has been established while

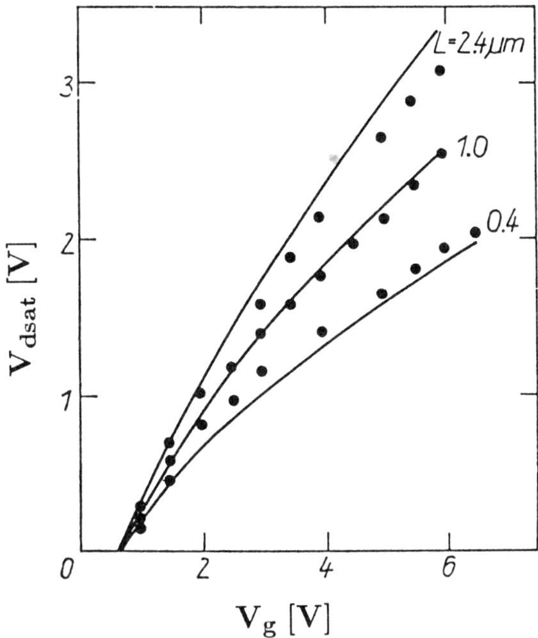

Figure 1.3: Experimental and theoretical variations of the saturation drain voltage V_{dsat} with gate voltage V_g for different channel lengths (parameters : $V_t = 0.6$ V, $\mu_0 = 500$ $cm^2/(Vs)$, $t_{ox} = 26$ nm, $v_{sat} = 6 \times 10^6$ cm/s, after [12]).

assuming μ_{eff} constant with gate voltage (or inversion charge). In the case where $\theta \neq 0$, the saturation drain voltage is no longer an explicit function of gate voltage [20]. V_{dsat} is given by the following implicit equation [20],

$$\theta^2[V_{dsat} + \frac{KV_{dsat}}{\theta} - (V_g - V_t)] = K[1 + \theta(V_g - V_t - V_{dsat})] \times$$

$$\ln\left[\frac{1 + \theta(V_g - V_t)}{1 + \theta(V_g - V_t - V_{dsat})}\right] \quad (1.29)$$

In Figure 1.4 are reported the variations of the drain saturation voltage V_{dsat} with the mobility attenuation factor θ obtained numerically using (1.28) and (1.29). Note that Equation 1.28 provides a good approximation for V_{dsat} until θ exceeds 0.1 V^{-1}. Above this value (*i.e.* for gate oxide thicknesses below 20 nm) the influence of θ on V_{dsat} is too high to be neglected and, thereby, Equation 1.29 becomes necessary. Nevertheless, the discrepancy between the two formulas does not exceed $\approx$ 20%. Another phenomenon which occurs in short-channel devices is the Drain Induced Barrier Lowering (DIBL). This effect is particularly important in the subthreshold region where the electrostatic

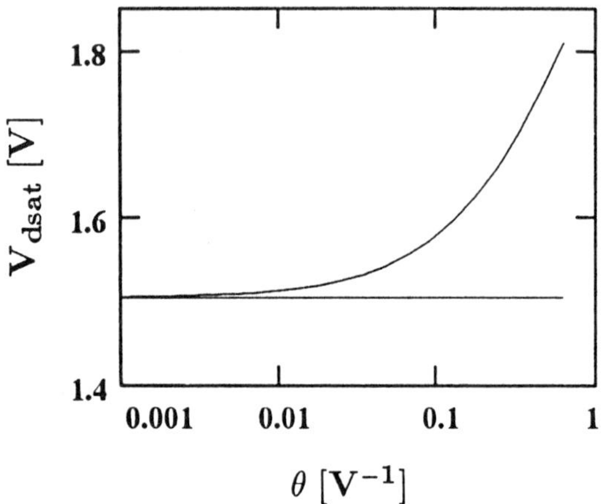

Figure 1.4: Variation of the saturation drain voltage V_{dsat} with the mobility reduction factor θ as obtained from Equations 1.28 and 1.29 (parameters : $\mu_0 = 500\ cm^2/(Vs)$, $t_{ox} = 20\ nm$, $v_{sat} = 6 \times 10^6\ cm/s$, $L = 0.3\ \mu m$, $V_g = 3\ V$, $V_t = 1\ V$).

screening is weak. At first order, the surface potential increase at the source, $\Delta\psi_s$, can be related to the drain bias as $\Delta\psi_s = BV_d$ where B is the DIBL coefficient given by [2,9] :

$$B = \frac{\epsilon_{si} t_{ox}}{\pi \epsilon_{ox} L} \tag{1.30}$$

where t_{ox} is the gate oxide thickness and ϵ_{ox} the oxide permittivity. The drain current in weak inversion is then given by an expression similar to Equation 1.21 in which I_{dsat} is replaced by $I_{dsat} \exp(qBV_d/kT)$ [9,21]. The DIBL effect not only alters the drain saturation current but also the output dynamic conductance even at low drain voltages [21]. In fact, for $V_d < kT/q$, the output conductance decreases exponentially with V_d as [21,22] :

$$\ln\left(\frac{g_d}{G_d}\right) \simeq -(A - 2B)\frac{qV_d}{kT} \tag{1.31}$$

where $G_d = g_d(0)$ is the ohmic conductance. The DIBL is not only effective in weak inversion but also affects the saturation performance of short channel devices in strong inversion. Indeed, the threshold voltage is reduced by an amount proportional to the drain voltage such that $V_t = V_{t0} - \lambda V_d$ with $\lambda = CB$. Since the drain current is a function of the drain voltage and the gate voltage drive $(V_g - V_t)$ from weak to strong inversion, it is easy to prove that the output conductance of the actual device can be expressed as [23] :

$$g_d = g_{d0} + \frac{\partial I_d}{\partial V_t}\frac{dV_t}{dV_d} = g_{d0} - \frac{\partial I_d}{\partial V_g}\frac{dV_t}{dV_d} = g_{d0} + \lambda g_m \tag{1.32}$$

where g_{d0} refers to the device output conductance in the absence of DIBL and g_m is the gate transconductance of the actual device. In the saturation region, g_{d0} cancels such that the output conductance reduces to [23] :

$$g_{dsat} = \lambda g_{msat} \tag{1.33}$$

where g_{msat} is the saturation transconductance. Equation 1.33 clearly indicates that the output conductance in saturation no longer vanishes as in long channel devices causing a degradation of the output characteristics of the device. Also of importance in the case of short channel devices is the channel length shortening which results from the extension at high drain voltage of the depletion layer near the drain. This phenomenon gives rise to a relative reduction of the effective channel of the form [24] :

$$\frac{\Delta L}{L} = \alpha \ln \left(1 + \frac{V_d - V_{dsat}}{\alpha V_p} \right) \tag{1.34}$$

where α and V_p are fitting parameters. Both the DIBL and the channel length shortening contribute to degrade the output conductance in the saturation region.

1.3   MOSFET Parameter Extraction Methods

The MOSFET parameters to be extracted can be classified into two categories: i) the technological parameters and ii) the electrical parameters. The former ones are essentially controlled by the design and fabrication processes. These are the gate oxide capacitance, the channel doping and the mask gate length and width L_{mask} and W_{mask}. The latter ones are determined by the electrical configuration of the device ; these are the effective channel length and width L and W, the source-drain series resistance R_{sd}, the threshold voltages V_t and V_{text}, the mobility parameters μ_0 and θ, the DIBL coefficients B and λ, the saturation drain voltage V_{dsat}, the saturation velocity v_{sat} and the fast interface trap density D_{it}.

The extraction of the static MOSFET parameters is in general performed using a set of transfer or output characteristics $I_d(V_g)$ and $I_d(V_d)$ collected in the ohmic and non linear regions of operation on devices of various geometries [25-28,31]. The ohmic MOSFET parameters can be obtained from the weak or strong inversion regions where certain approximations hold (see section 1.2). The parameters of interest which can be deduced from the strong inversion region are the charge threshold voltage V_t, the extrapolated threshold voltage V_{text}, the low field mobility μ_0, the mobility attenuation factor θ, the inversion charge to depletion charge weighting factor η, the source-drain series resistance R_{sd}, the effective channel doping N_A and the effective channel length and width of the devices L and W. From the weak inversion characteristics, one can deduce the channel doping, the interface trap density D_{it} and the DIBL coefficient B.

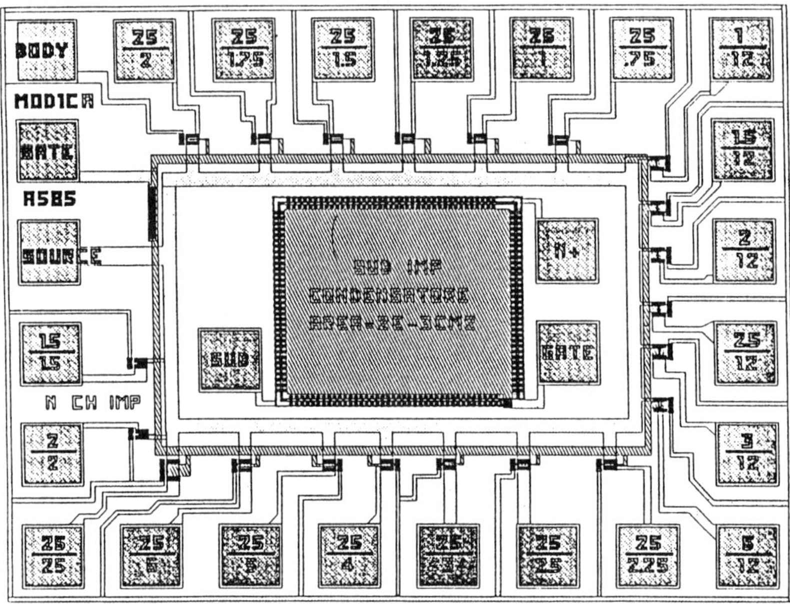

Figure 1.5: Example of a typical test mask used for the MOSFET parameter extraction with devices of various gate lengths and gate widths (courtesy of SGS-Thomson).

From the non linear operation region, other relevant MOSFET parameters can be extracted such as the saturation drain voltage V_{dsat}, the DIBL coefficient λ or the saturation velocity v_{sat}.

For this purpose, a typical test mask including a set of MOSFETs with various gate lengths and fixed gate width, and vice versa, as well as a large area MOS capacitor has to be available on the chip of interest. Figure 1.5 shows an example of such typical test mask used for MOSFET parameter extraction.

The large area MOS capacitor is used for the measurement of the gate oxide capacitance per unit area C_{ox}. This can be done using a conventional LCR meter while biasing the gate in strong accumulation so that the semiconductor surface may behave as a metallic electrode. Then sets of data for the transfer and output characteristics are collected for different transistors with various geometries.

1.3.1 Parameter Extraction in the Ohmic Regime

Charge and Extrapolated Threshold Voltages

A useful function for the extraction of the ohmic MOSFET parameters is based on the combination of the drain current and transconductance relations 1.14

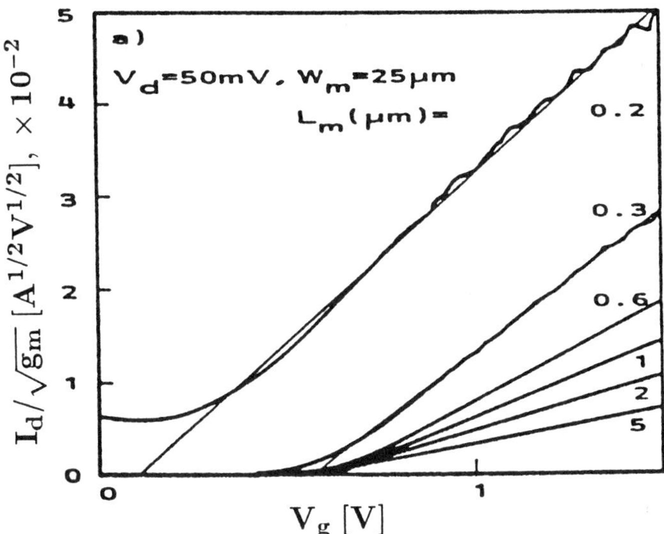

Figure 1.6: Typical experimental $Y(V_g)$ characteristics as obtained on n-channel MOS devices from 0.35 μm CMOS technology used for the extraction of the charge threshold voltage V_t and low field mobility μ_0 (courtesy of O. Roux).

and 1.15 as [29] :

$$Y(V_g) = \frac{I_d}{\sqrt{g_m}} = \sqrt{G_m V_d}(V_g - V_t) \tag{1.35}$$

where $G_m = WC_{ox}\mu_0/L$ is the transconductance parameter.

Equation 1.35 clearly indicates that if the mobility law of Equation 1.9 used in Equations 1.14 and 1.15 is correct, the Y function has to vary linearly with gate voltage with the slope and x-axis intercept providing the parameter G_m and the charge threshold voltage V_t, respectively. It should be noted that the Y function is independent of the series resistance effects as can be demonstrated experimentally by adding external resistors in series with the drain or source terminals [30]. Moreover, it is worth mentioning that the extrapolated threshold voltage V_{text} which can be obtained from the inflection point coordinates using (1.16) is not equivalent to the charge threshold voltage V_t deduced from the $Y(V_g)$ function. Unlike V_t, the extrapolated threshold voltage V_{text} is affected by the mobility attenuation factor and the series resistance. V_{text} can only be taken as a rough approximation of the charge threshold voltage.

A typical example of $Y(V_g)$ characteristics illustrating the linear variations with gate voltage is shown in Figure 1.6 for a 0.35 μm CMOS technology. Note the excellent optimization of the devices down to 0.3 μm gate length.

Effective Dimensions, Mobility and Series Resistance

After having collected a set of G_m data for all the gate lengths and widths available, the effective channel dimensions can be determined. At this point, it is worth pointing out that G_m is proportional to the effective channel width W and inversely proportional to the effective channel length L such that [31],

$$G_m = \frac{\mu_0 C_{ox} W}{L} = \frac{\mu_0 C_{ox}}{L}(W_{mask} - \Delta W) \qquad (1.36)$$

and

$$\frac{1}{G_m} = \frac{L}{\mu_0 C_{ox} W} = \frac{1}{\mu_0 C_{ox} W}(L_{mask} - \Delta L) \qquad (1.37)$$

where ΔL and ΔW are the channel length and width reductions which result from lateral diffusion and/or over etching during the gate definition process.

Provided the low field mobility μ_0 is independent of channel length and width, the plot of G_m as a function of gate width for a fixed gate length should give a straight line with the x-axis intercept providing the channel width reduction ΔW as in Figure 1.7(a). Likewise, the plot of $1/G_m$ versus gate length may also be linear, giving access to the channel length reduction ΔL as shown in Figure 1.7(b).

Knowing the effective channel length and width for each device enables the low field mobility μ_0 to be deduced from the G_m data as $\mu_0 = L G_m/(W C_{ox})$.

The mobility attenuation factor θ^* for each device can then be obtained from the plot of $1/\sqrt{g_m}$ as a function of gate voltage. In fact, one has from (1.15),

$$\frac{1}{\sqrt{g_m}} = \frac{[1 + \theta^*(V_g - V_t)]}{\sqrt{G_m V_d}}. \qquad (1.38)$$

The slope of this quantity plotted with gate voltage provides the extrinsic mobility attenuation factor θ^* (see Figure 1.8). Since $\theta^* = \theta + G_m R_{sd}$, a further plot of the extrinsic mobility attenuation factor θ^* versus G_m for various gate lengths (with fixed gate width) allows one to deduce the intrinsic mobility attenuation factor θ and the source-drain series resistance R_{sd} from the y-axis intercept and slope, respectively [32] (see Figure 1.9).

Alternative methods for the extraction of ΔL and R_{sd} have been proposed which are based on the fact that the channel sheet resistivity $\rho_{ch} = 1/(Q_i \mu_{eff})$ is only a function of the gate voltage drive $(V_g - V_t)$ such that the total resistance reads [33,34],

$$R_{tot} = \frac{L \rho_{ch}}{W} + R_{sd} = \frac{(L_{mask} - \Delta L)\rho_{ch}}{W} + R_{sd} \qquad (1.39)$$

Therefore, a plot of R_{tot} as a function of gate length should be linear for a given gate voltage. The common intersection point of the observed straight lines obtained for various gate voltages (or gate voltage drives) provides the channel length reduction ΔL and the series resistance as shown by Figure 1.10.

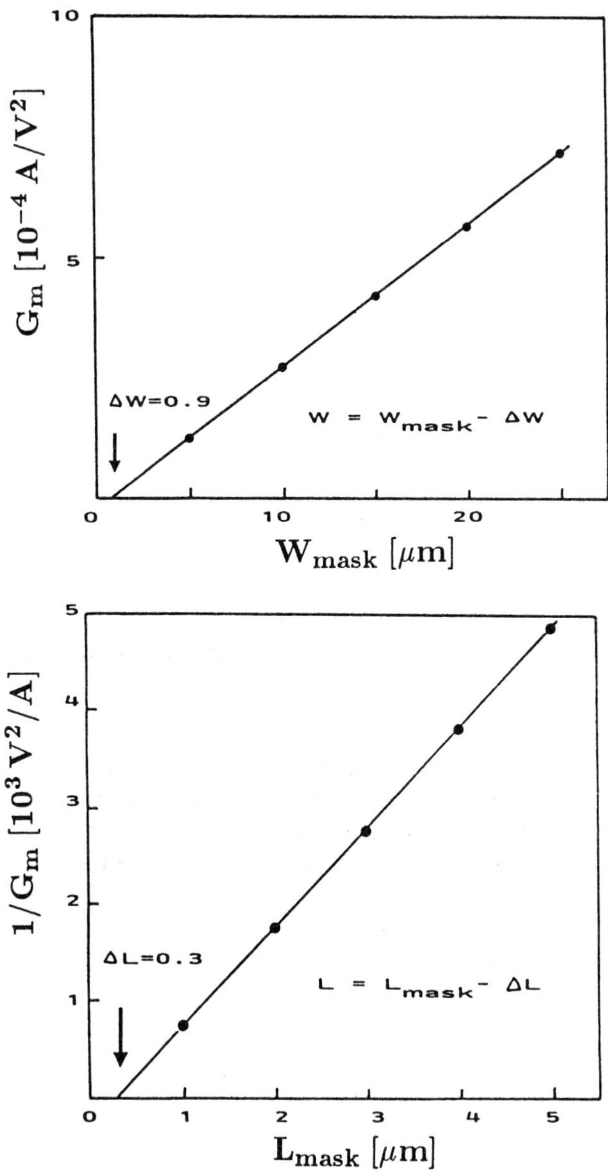

Figure 1.7: Variations of G_m with gate width W_{mask} (a) and $1/G_m$ with gate length L_{mask} (b) illustrating the extraction of the channel reduction width and length (after [31]).

16

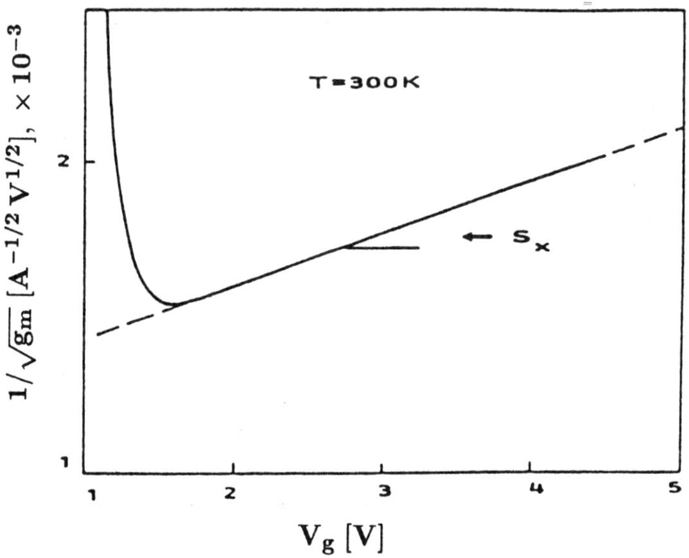

Figure 1.8: Typical variation of $g_m^{-1/2}$ with gate voltage V_g used with Equation 1.27 for the evaluation of the mobility attenuation factor θ (courtesy of A. Emrani).

While employing this method, one can be confronted with the fact that no common intersection point exists for various gate voltages. In such a case (see Figure 1.11), it is admitted that each intersection point corresponding to a couple of gate voltage values, V_{g1} and V_{g2}, does provide the channel length reduction and series resistance for the mean gate voltage (or gate voltage drive). As a result, it is possible to evaluate the dependence of the channel length reduction and series resistance with gate voltage (or gate voltage drive) [33,34].

Channel Doping

The effective channel doping of the MOS transistor can be extracted from the sensitivity curve of the charge threshold voltage V_t to the substrate voltage V_b. In fact, it is easy to demonstrate from (1.4) that [31,35] :

$$\frac{\Delta V_t}{\Delta V_b} = -\frac{C_d}{C_{ox}} \tag{1.40}$$

where C_d is the depletion capacitance at threshold : $C_d = \sqrt{q\epsilon_{si}N_A/4\Phi_f}$.

Knowing the depletion capacitance, it is then easy to deduce from the abacus of Figure 1.12 or from numerical resolution of (1.40) the value of the corresponding doping level.

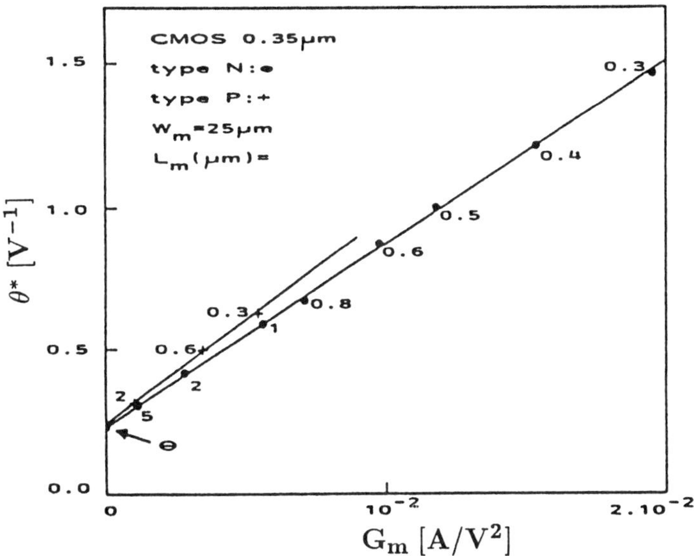

Figure 1.9: Plot of the extrinsic mobility attenuation factor θ^* versus G_m used for the extraction of the intrinsic mobility attenuation factor θ and source drain series resistance R_{sd} (courtesy of O. Roux).

Equation 1.40 can in principle be applied to extract the doping profile in the channel. To this end, one has to measure the dependence of the charge threshold voltage with the bulk bias $V_t(V_b)$, and, differentiate it with respect to V_b. Making use of (1.40), it is then possible to plot $1/C_d^2$ as a function of bulk bias V_b, and, extract the doping level from the local slope for each bulk bias (*i.e.* each depletion layer width), as in the case of MOS capacitor [36]. In such situation, the local slope is related to the doping level at the width $W(V_b)$ by [36],

$$\frac{d}{dV_b}\left(\frac{1}{C_d^2}\right) = -\frac{2}{q\epsilon_{si}N_A(W)} \tag{1.41}$$

where $W(V_b) = \epsilon_{si}/C_d(V_b)$.

Interface Trap Density

The interface trap density D_{it} entering Equation 1.13 via the capacitive ratio A can be evaluated using the subthreshold slope $S = g_m/I_d = qA/kT$. In order to extract C_{it} (and consequently D_{it}), it is necessary to know C_{ox} and the depletion capacitance in weak inversion C_d^w. For this, one can use the value of the channel doping previously determined and evaluate $C_d^w = \sqrt{q\epsilon_{si}N_A/3\Phi_f}$.

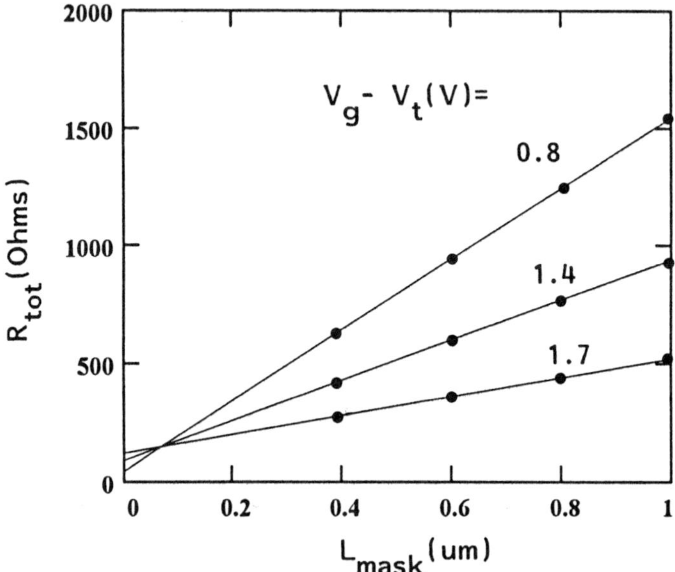

Figure 1.10: Plot of the total resistance R_{tot} versus gate length L_{mask} used for the simultaneous extraction of the channel reduction length ΔL and series resistance in the case of a common intersection point.

Then the interface trap density can be deduced from,

$$D_{it} = \frac{C_{ox}}{q} \left(\frac{q}{SkT} - 1 - \frac{C_d}{C_{ox}} \right) \tag{1.42}$$

where D_{it} being in $eV^{-1}cm^{-2}$.

It should be noted that the subthreshold slope technique is only appropriate when the interface trap density is relatively high ($> 5 \times 10^{10} \sim 10^{11}\ eV^{-1}cm^{-2}$). For low interface trap densities, it becomes quite inaccurate, since D_{it} has to be extracted from a difference of capacitances. Assuming that $C_d + C_{ox}$ is known with a precision of 10%, a typical value of $10^{-7}\ F/cm^2$ leads to a minimum detectable interface trap density of about $6 \times 10^{10}\ eV^{-1}cm^{-2}$. This feature proves that the subthreshold slope technique is not adequate for the evaluation of the interface trap density in virgin MOS devices fabricated using state-of-the-art technology where D_{it} lies in the range $5 \times 10^9 - 10^{10}\ eV^{-1}cm^{-2}$. Nevertheless, this technique can be very useful for a fast characterization of stressed devices [37]. Figure 1.13(a) gives an example of subthreshold characteristics after constant current stress carried out on n-channel devices and the corresponding interface trap density as a function of the injection dose (Figure 1.13(b)). The comparison to data obtained with charge pumping measurements clearly

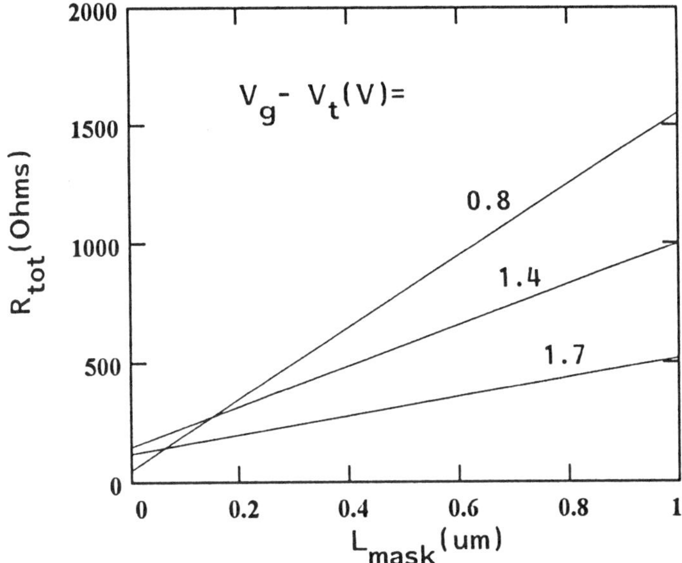

Figure 1.11: Plot of the total resistance R_{tot} versus gate length L_{mask} showing its possible use for the simultaneous extraction of the channel reduction length ΔL and series resistance in the case of gate voltage dependent intersection point.

demonstrates that the subthreshold technique suffers from inaccuracy below $10^{11}\ eV^{-1}cm^{-2}$.

Mobility Weighting Factor

The extraction of the mobility weighting factor η entering the effective mobility law of Equations 1.9 and 1.10 can be carried out by making use of the body-to-gate transconductance ratio, g_b/g_m, plotted as a function of gate voltage [38]. As a matter of fact, it is possible to show from Equations 1.14 and 1.15 that, in strong inversion, one has [38],

$$\frac{g_b}{g_m} = \frac{C_d}{C_{ox}} \left[1 + \frac{\theta}{\eta}(V_g - V_t) \right] \tag{1.43}$$

Figure 1.14 shows typical characteristics of the body-to-gate transconductance ratio as a function of gate voltage which illustrates the validity of Equation 1.43. The slope of the linear part above threshold enables the quantity η to be deduced while knowing the mobility attenuation factor θ. It should be mentioned that these plots also allow the determination of the ratio C_d/C_{ox}

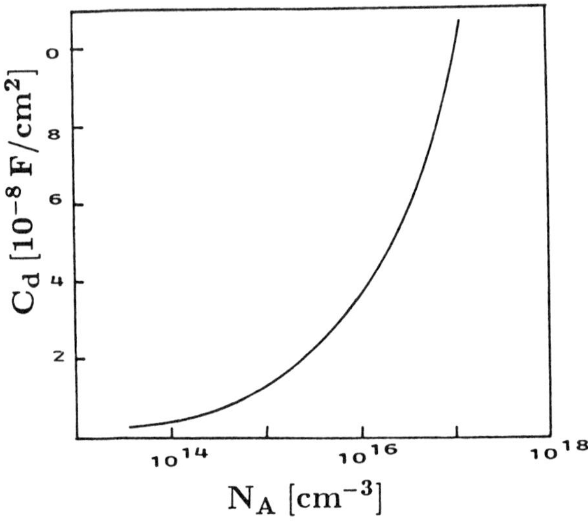

Figure 1.12: Abacus giving the variation of the depletion capacitance C_d at threshold with substrate doping N_A.

as indicated on the figure. Using this technique, typical values of η have been extracted around 1/2 for n-channel devices, and 1/3 for p-channel devices in good agreement with data obtained using split-capacitance techniques [15]. The advantage of the above method is that it is applicable on large as well as on small area MOS devices, since it is based on drain current measurements, and not on capacitive ones. A direct η extraction technique based on the derivatives of the effective mobility with respect to gate voltage and bulk bias has recently been proposed. This technique allows one to check whether or not η is constant with gate voltage [39].

1.3.2 Parameter Extraction in Saturation

The principal parameters which characterize the non linear MOSFET operation are the saturation drain voltage V_{dsat}, the saturation velocity v_{sat} and the DIBL coefficient B (or λ). Moreover, one can also obtain information about the interface trap density while using the dynamic conductance in weak inversion (see Equation 1.31 [22]).

Saturation Drain Voltage

A simple method for the extraction of the saturation drain voltage relies naturally on a rapid analysis of the drain current output characteristics $I_d(V_d)$. Thus, V_{dsat} is defined as the point where the drain current saturates. The

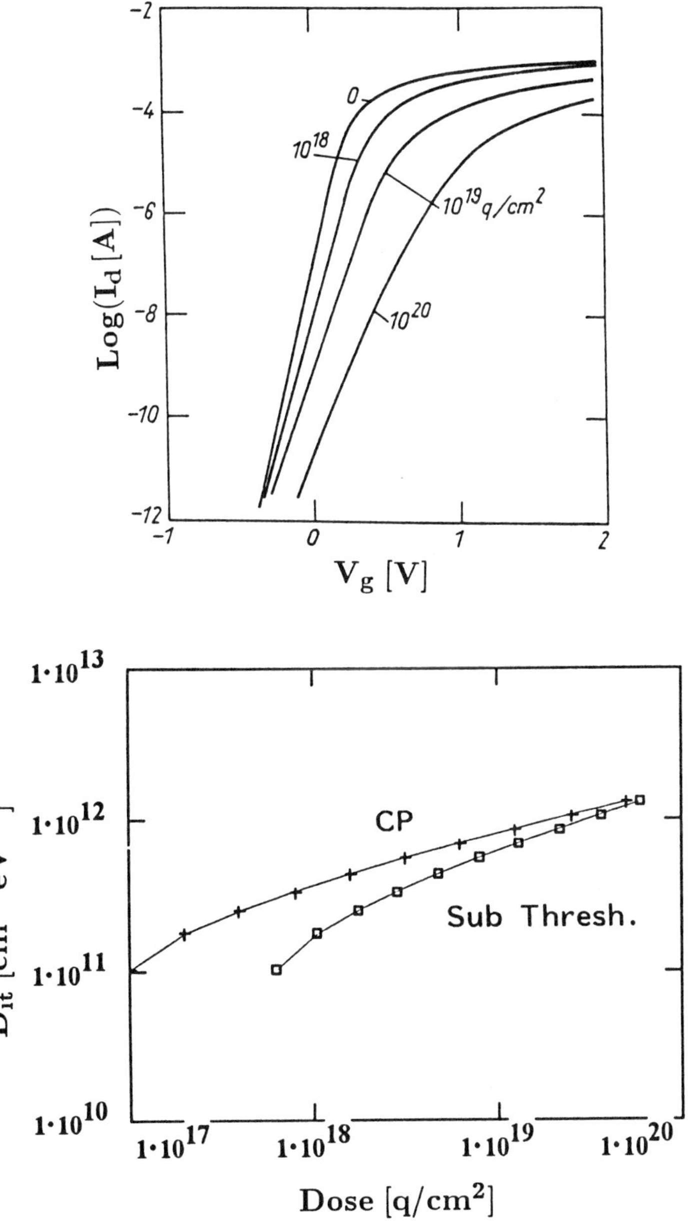

Figure 1.13: Typical subthreshold drain current characteristics as obtained on MOS devices after several uniform gate stress (a) and corresponding evolution of the interface trap density D_{it} with the injection dose (b) obtained from subthreshold technique and charge pumping measurements (after [37]).

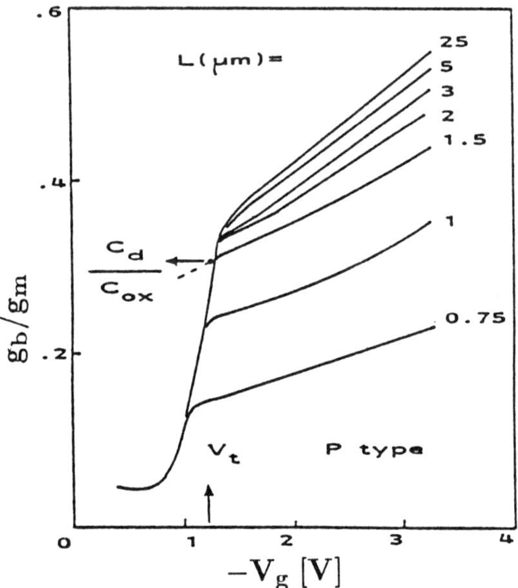

Figure 1.14: Typical variations of the body-to-gate transconductance ratio g_b/g_m with gate voltage V_g used for the extraction of the mobility parameter η (after [38]).

problem with this method is that the output characteristics of an actual device never saturate ideally, especially in the case of short channel devices. Therefore several criteria have been proposed in order to overcome this difficulty.

One of these methods is based on the use of the output conductance characteristics [40]. In such a case, the saturation drain voltage is defined at the point where the normalized output conductance g_d/G_d takes a fixed value (e.g. 0.1 or 0.01). This method has been successfully applied to MOS devices with channel lengths down to 0.4 μm [40]. However, an arbitrary value has to be chosen for the reduction rate of g_d/G_d.

Another method proposed for the determination of V_{dsat} relies on the second derivative of the drain current characteristics using the function [41],

$$G(V_d) = g_d \frac{d}{dV_d} \left(\frac{1}{g_d} \right) \tag{1.44}$$

The function $G(V_d)$ exhibits a bell shape behavior with drain voltage with a maximum located near the saturation point V_{dsat}. Using this method enables satisfactory saturation drain voltage values to be obtained as a function of gate voltage as shown in Figure 1.15.

Alternative methods for the extraction of the saturation drain voltage make

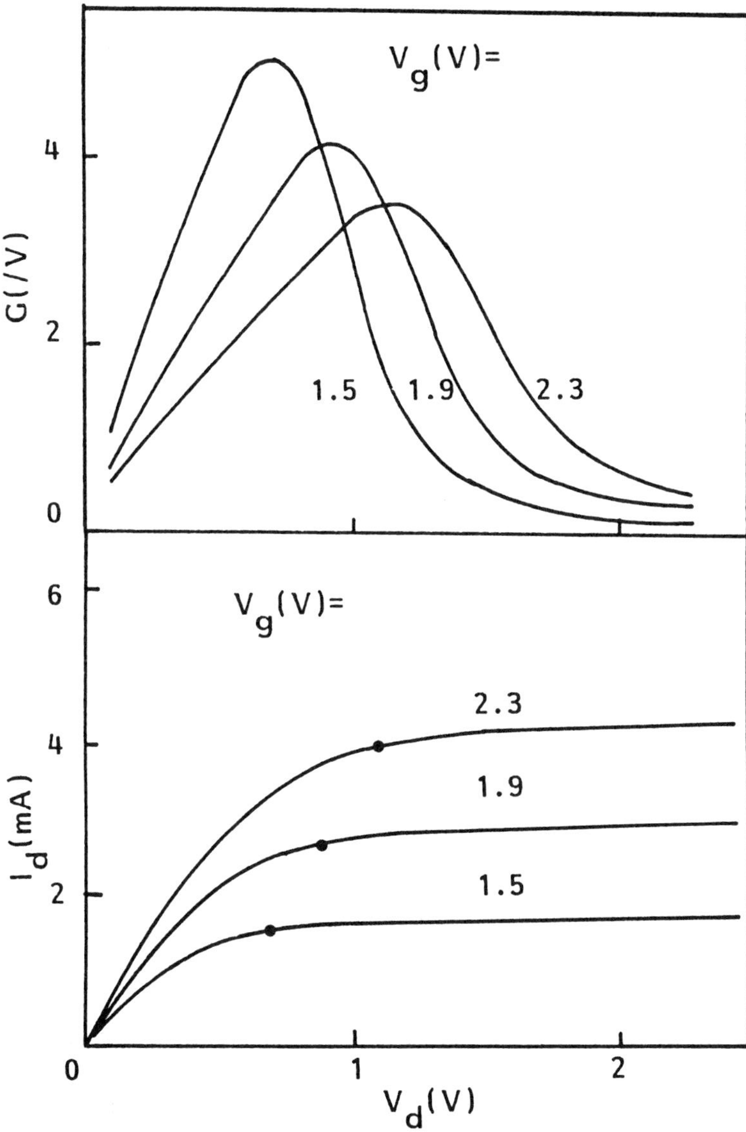

Figure 1.15: (a) Variations of the function $G(V_d)$ (Equation 1.43) and (b) corresponding saturation drain voltages on $I_d(V_d)$ characteristics (courtesy of H. Brut).

24

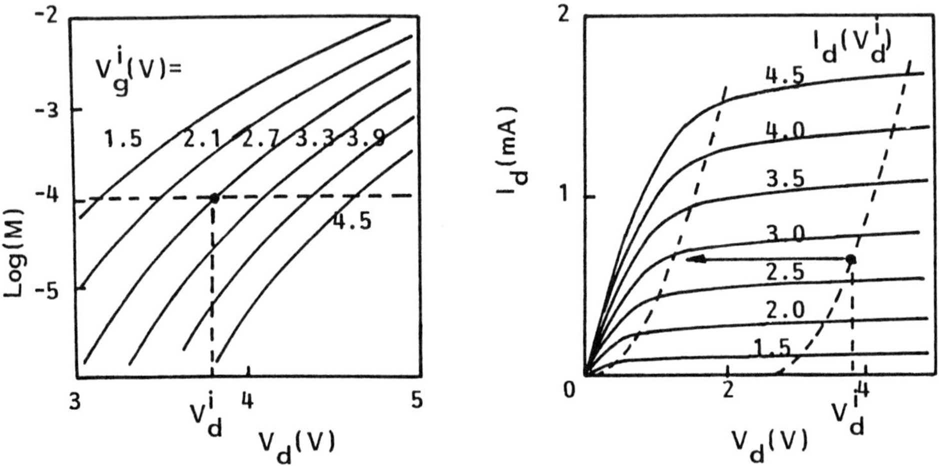

Figure 1.16: (a) Variations of the multiplication factor M with drain voltage for various gate voltages and (b) corresponding drain current output characteristics illustrating the extraction of the saturation drain voltage based on Equation 1.45 (after [42]).

use of the dependence of the multiplication factor on the saturation drain voltage [42,43]. Indeed, the impact ionization substrate current is generally a function of V_{dsat} such that [42],

$$\frac{I_{sub}}{I_d} = M(V_d - V_{dsat}) \tag{1.45}$$

$$= A(V_d - V_{dsat})\exp\left(-\frac{B}{V_d - V_{dsat}}\right) \tag{1.46}$$

where A and B are pre- and exponential impact ionization coefficients.

Measuring for each gate voltage V_g^i the loci in drain voltage V_d^i for a constant multiplication factor (e.g. $I_{sub}/I_d = 10^{-4}$) enables the corresponding saturation drain voltage to be related to V_d^i as,

$$V_{dsat}^i = V_d^i - M^{-1}(10^{-4}) \tag{1.47}$$

Finally, assuming that $M^{-1}(10^{-4})$ is constant with gate voltage and that V_{dsat} cancels below threshold allows one to evaluate the unknown constant $M^{-1}(10^{-4})$, and, in turn, to obtain the actual value of the saturation drain voltage from Equation 1.45 (see Figure 1.16).

A similar method uses the sensitivity of the multiplication factor to the drain and gate voltages [43]. As a matter of fact, assuming again that M is only a function of $V_d - V_{dsat}$ implies that,

$$\frac{dV_{dsat}}{dV_g} = -\frac{dM/dV_g}{dM/dV_d} \tag{1.48}$$

Therefore, the measurements of the sensitivities of M with respect to V_g and V_d for given drain and gate voltages provide in a simple way the derivative of the saturation drain voltage with respect to the gate voltage. A further integration over gate voltage can then be used to obtain the variation of the saturation drain voltage with gate voltage as [43],

$$V_{dsat} = -\int_{V_t}^{V_g} \frac{dM/dV_g}{dM/dV_d} dV_g \tag{1.49}$$

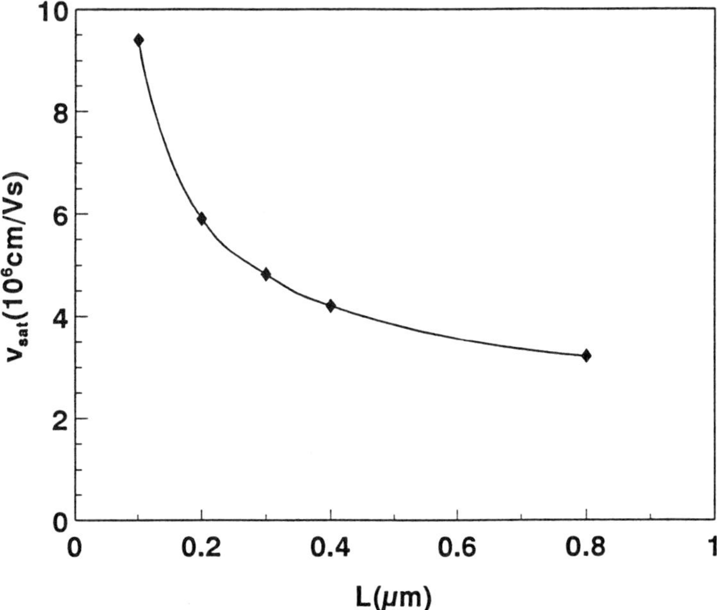

Figure 1.17: Variation of the saturation velocity v_{sat} with channel length as obtained using Equations 1.50 and 1.51 illustrating the onset of velocity overshoot (after [47]).

Saturation Velocity

Generally, the extraction of the saturation velocity is conducted using the output drain current characteristics assuming that the drift velocity of the carriers

26

at the drain edge attains saturation for sufficiently large drain voltages [13,14]. Therefore, the saturation velocity at strong inversion can be obtained as [44],

$$v_{sat} = \frac{I_{dsat}}{WC_{ox}(V_g - V_t - V_{dsat})} \tag{1.50}$$

Similarly, assuming that v_{sat} is independent of gate voltage, the saturation velocity can also be extracted from the saturation transconductance from [44],

$$v_{sat} = \frac{g_{msat}}{WC_{ox}\left(1 - dV_{dsat}/dV_g\right)} \tag{1.51}$$

The interest of the latter formula is that it does not require the determination of the threshold voltage and uses the sensitivity of the saturation drain voltage with gate voltage (see Equation 1.48).

Figure 1.17 shows typical saturation velocity data obtained using Equations 1.50 and 1.51 on short channel MOS devices and which illustrate the onset of the so-called velocity overshoot phenomenon when the transit time becomes small enough to get non stationary transport [45]. It should also be noted that the saturation velocity data obtained directly from a MOSFET using these methods compare favorably with those deduced from resistive gated MOS devices under uniform electric field condition [46,47].

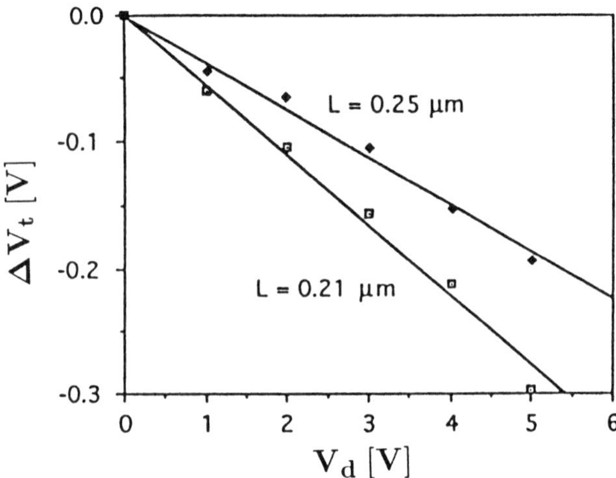

Figure 1.18: Dependence of the threshold voltage on drain voltage observed in short channel devices where DIBL occurs (after [48]).

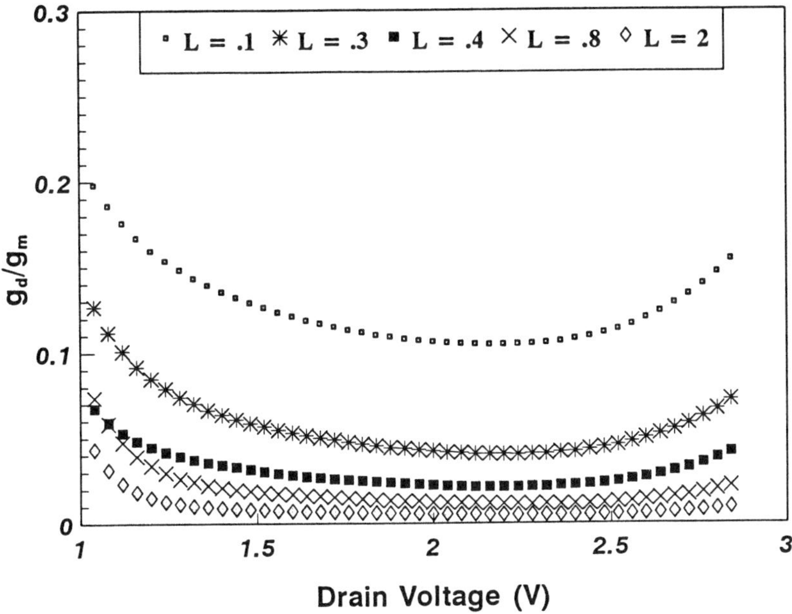

Figure 1.19: Typical variations of the ratio g_d/g_m with drain voltage V_d for various gate lengths (after [23]).

DIBL Coefficient

The extraction of the DIBL coefficient can be carried out from the output drain current characteristics of the MOSFET operated either in weak inversion or above threshold.

The $I_d(V_d)$ curves recorded in weak inversion generally exhibit an exponential behavior in saturation. The logarithmic slope of the linear part of the characteristics allows one to deduce the DIBL coefficient B from [9],

$$B = \frac{kT}{q} \frac{d}{dV_d} (\ln I_d) \tag{1.52}$$

The DIBL coefficient can also be determined from the strong inversion characteristics in saturation. It is possible to measure the extrapolated threshold voltage as a function of drain bias, and if a straight line is observed at high V_d values, one can obtain the DIBL coefficient λ from the slope. An example of such a behavior for the threshold voltage is given in Figure 1.18 [48].

Recently, another method for the extraction of the DIBL parameter λ has been proposed based on the use of the dynamic conductance expression of Equation 1.33 [23], where the ratio g_d/g_m is plotted as a function of drain voltage for a fixed gate voltage. Figure 1.19 gives a typical example of g_d/g_m versus

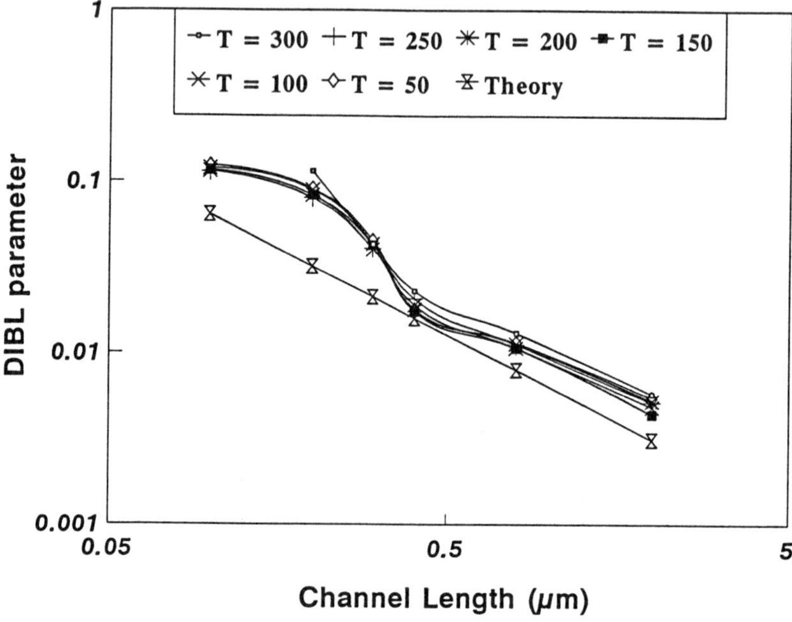

Figure 1.20: Experimental variation of the DIBL parameter λ with gate length L as obtained from the g_d/g_m method and theoretical plot using Equation 1.30 (after [23]).

V_d as obtained for various gate lengths ranging between 0.1 and 2 μm for a fixed gate voltage. The DIBL parameter λ for each gate length is then taken at the minimum of the plateau regions. The corresponding values of λ deduced by this method are displayed in Figure 1.20 for various gate lengths. Note the approximate $1/L$ dependence of λ in good agreement with the simple model of Equation 1.30.

Unlike the previous procedures, this method can be applied indifferently from weak to strong inversion and therefore allows one to determine the gate voltage dependence of the DIBL coefficient λ. Moreover, it has been used successfully to study the influence of temperature on the DIBL effect [23].

1.4   Measuring Techniques

The MOSFET drain current characteristics are generally measured using commercially available source measure units (SMU) such as Model K236-238 from Keithley Instruments Inc (USA) or Model HP4142-4155 from Hewlett-Packard (USA). These systems provide in a compact form both the voltage sources for biasing the device terminals and current meters for measuring the current flow-

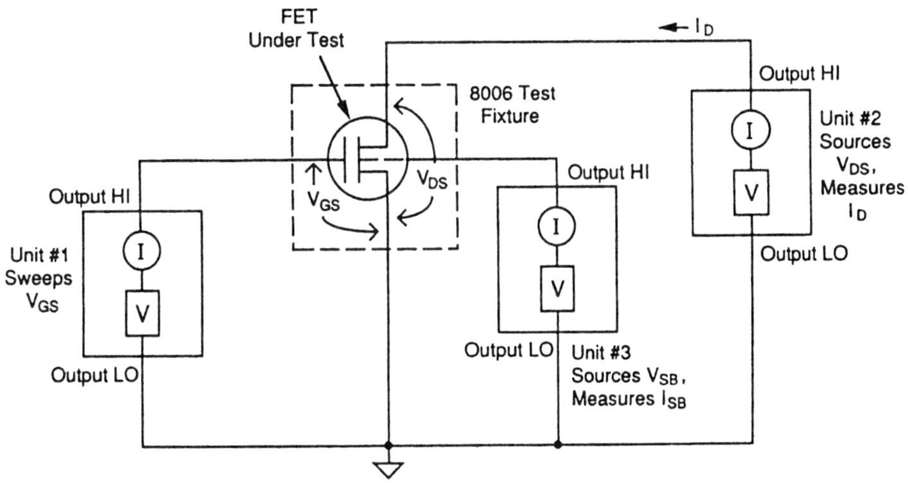

Figure 1.21: Source Measure Unit configuration currently used for the measurement of three terminal (MOSFET) characteristics (after [49]).

ing in a specific connection lead.

An SMU is therefore composed basically of a voltage source and a current meter placed in series. In order to measure the transfer and output characteristics of a three terminal device as a MOSFET, it is necessary to employ two SMU's to supply the gate and drain voltages and recording the drain current while grounding the source and substrate. If one also needs to bias the substrate terminal and/or measure the substrate current, a third SMU is required. A typical three terminal FET measuring configuration is shown in Figure 1.21.

The standard specifications for such source measure units range typically between 100 μV and 100 V for the voltage sources, and between 10 fA and 100 mA for the current meters [49,50]. The source measure units can generally be used in static mode or in sweep mode. In static operation, a constant DC voltage is applied to the output terminal. In this case, a constant voltage pulse with given hold time is sent to the output. Concurrently, the current meter readings are taken using various integration times and average numbers. The accuracy of the voltage is typically less than 0.04%. The 4 or 5 digit current meter resolution lies in a range of $10^{-4} - 10^{-5}$ with a 0.04% accuracy.

Both coaxial and triaxial connection leads can be employed with these commercial systems in order to minimize the leakage current between the ouput high and ouput low lead connections [49,50]. The source measure unit can provide a driven guarding which surround the ouput high lead connection. It is recommended to use the guard shield for measuring current level below $\approx 1 \mu A$. The source measure units can be operated in stand alone mode using the sweep possibilities and data buffering. More conveniently, they can be controlled by a desk top computer with a dedicated measurement program which command adequately the SMU through the 488-IEEE bus.

1.5   Summary and Conclusion

The basic equations necessary for the modeling of the static MOSFET operation have been presented. The relevant charges, potentials and drain current equations of the MOS transistor have been established. Moreover, the principal MOSFET parameters which govern the ohmic and non linear operation regions have been defined. The various analytical approximations used in the strong and weak inversion regions have also been discussed. Finally, the refinements introduced for the modeling of submicronic phenomena such as short channel effects, Drain Induced Barrier Lowering or channel length shortening have also been addressed. Then, the methods currently used for the extraction and the characterization of the main MOSFET parameters have been exposed. The concepts of charge and extrapolated threshold voltages have been clarified. The relevant mobility coefficients such as the low field mobility and the mobility attenuation factor have been explained. The notion of effective electrical channel length and channel width have also been presented. Eventually, the basic concepts and experimental means required for the measurements of the static MOSFET characteristics have been briefly described.

Bibliography

[1] G. Merckel, J. Borel, N. Cupcea, *IEEE Trans Electron Devices*, 19, 681 (1972).

[2] R. Swansson, J. Meindl, Proc. *IEEE Int Solid State Circuits Conf.*, 110 (1975).

[3] J. R. Brews, *Sol State Electron*, 21, 345 (1978).

[4] P. Muls, G. Declerck, R. Van Overstraeten, *Advances in Electronics and Electron Physics*, 47, 197 (1978).

[5] M. White, F. Van de Wielc, J. P. Lambot, *IEEE Trans Electron Devices*, 27, 899 (1980).

[6] L. Akers, J. J. Sanchez, *Sol State Electron*, 25, 621 (1982).

[7] T. Yamaguchi, S. Morimoto, *IEEE Trans Electron Devices*, 30, 559 (1983).

[8] P. Guebels, F. Van de Wiele, *Sol State Electron*, 26, 267 (1983).

[9] T. Grottjohn, B. Hoefflinger, *IEEE Trans Electron Devices*, 31, 234 (1984).

[10] G. Ghibaudo, *Phys. Stat. Solidi (a)*, 95, 323 (1986).

[11] A. Hiroki, S. Odamaka, S. K. Ohe, H. Esaki, *IEEE Electron Devices Letters*, 8, 231 (1987).

[12] G. Ghibaudo, *Phys. Stat. Sol (a)*, 113, 223 (1989).

[13] S. M. Sze, *Physics of Semiconductor Devices*, Wiley, New -York, 1981.

[14] R.S. Muller and T.I. Kamins, *Device Electronics for Integrated Circuits* John Wiley, New-York, 1986.

[15] C.L. Huang and G. Gildenblat, *IEEE Trans Electron Devices*, ED-37, 1289 (1990).

[16] V.M. Agostinelli, H. Shin, A.F. Tasch, *IEEE Trans Electron Devices*, ED-38, 151 (1991).

[17] A. Emrani, F. Balestra, G. Ghibaudo, *IEEE Trans Electron Devices*, ED-40, 564 (1993).

[18] G. Merckel, *Sol. State Electron*, 23, 1207 (1980).

[19] Z.H. Liu, C. Hu, J.H. Huang, T.Y. Chan, M.C. Jeng, P.K. Ko, Y.C. Cheng, *IEEE Trans Electron Devices*, ED-40, 86 (1993).

[20] G. Pellegrini and R.L. Anderson, *J. Appl. Phys.*, 72, 3606 (1992).

[21] G. Ghibaudo, B. Cabon, *Electronics Letters*, 22, 1010 (1986).

[22] G. Ghibaudo, B. Cabon, *Sol State Electron*, 30, 1049 (1987).

[23] W. Fikry, G. Ghibaudo, M. Dutoit, *Electronics Letters*, 30, 911 (1994).

[24] F.M. Klaassen, *Proc. ESSDERC 90*, Nottingham, UK, 1990, Eds. W. Eccleston and P.J. Rosser (Adam Hilger, Bristyol 1990) p.181.

[25] H.S. Wong, M.H. White, T.J. Krutsick, R.V. Booth, *Sol. State Electron*, 30, 953 (1987).

[26] T.J. Krutsick, M.H. White, H.S. Wong, R.V. Booth, *IEEE Trans Electron Devices*, ED-34, 1676 (1987).

[27] P.R. Karlsson and K.O. Jeppson, *IEEE Trans Electron Devices*, ED-39, 2070 (1992).

[28] C. C. McAndrew and P.A. Layman, *IEEE Trans Electron Devices*, ED-39, 2298 (1992).

[29] G. Ghibaudo, *Electronics Letters*, 24, 543 (1988).

[30] I.M. Hafez, F. Balestra, G. Ghibaudo, *J. Appl. Phys.*, 68, 3694 (1990).

[31] D. Bauza and G. Ghibaudo, *Microelectronics Journal*, 25, 41 (1994).

[32] A. Emrani, G. Ghibaudo, F. Balestra, *Electronics Letters*, 29, 786 (1993).

[33] M. Ida and C. Kita, *Proc. of IEEE Int. Conf. on Microelectronics Test Structures*, March 1990.

[34] S. S. Chung and J.S. Lee, *IEEE Trans Electron Devices*, ED-40, 1709 (1993).

[35] J.R. Brews, *Physics of MOS transistors*, in Applied Solid State Science, Supp. 2A (Academic Press, 1981) p. 1.

[36] E. H. Nicollian and J.R. Brews, *MOS (Metal Oxide Semiconductor) Physics and Technology* John Wiley, New York, (1982).

[37] C. Nguyen-Duc, G. Ghibaudo, F. Balestra, *Phys. Stat. Sol (a)*, 126, 553 (1991).

[38] A. Emrani, G. Ghibaudo, F. Balestra, *Electronics Letters*, 27, 467 (1991).

[39] A. Emrani, G. Ghibaudo, F. Balestra, *Sol. State Electron*, 37, 111 (1993).

[40] G. Ghibaudo, *Phys Stat Solidi (a)*, 99, K149 (1987).

[41] W.Y. Jang, C.Y. Wu, H.J. Wu, *Sol. Stat. Electron*, 31, 1421 (1988).

[42] T.Y. Chan, P.K. Ko, C. Hu, *IEEE Electron Device Letters*, EDL-6, 551 (1985).

[43] K. Rais, PhD Thesis, University El Jadida (July 1994).

[44] C.G. Sodini, P.K. Ko, J.L. Moll, *IEEE Trans Electron Devices*, ED-31, 1386 (1984).

[45] G. Shahidi, D. Antionadis, H. Smith, *Electron Device Letters*, EDL-9, 94 (1988).

[46] A. Modelli and S. Manzini, *Sol. Stat. Electron*, 31, 99 (1988).

[47] K. Rais, G. Ghibaudo, F. Balestra, M. Dutoit, *Proc. 1st European Workshop on Low Temperature Electronics (WOLTE 1)*, Eds. G. Ghibaudo and F. Balestra, J. Phys. IV, C6, June 1994, p. 19.

[48] J.E. Chung, M.C. Jeng, J.E. Moon, P.K. Ko, C.Hu, *IEEE Trans Electron Devices*, ED-38, 545 (1991).

[49] Operating manual of model K236-238, Keithley Instruments Inc., Cleveland, USA, 1989.

[50] Operating manual of model HP4145, Hewlett-Packard Inc., Palo Alto, USA, 1990.

Small Signal Characterization of VLSI MOSFETs

Hisham Haddara
Ain Shams University
Cairo, Egypt

2.1 Introduction

The performance of an MOS transistor is determined to a great extent by the quality of its $Si - SiO_2$ interface. During the early stages of development of MOS technology, the MOS capacitor has been used as the principal tool for investigating and understanding the properties of the MOS system [1-8]. In fact, much of the present knowledge of the properties of this system has been acquired by small signal admittance measurements on MOS capacitors. As pointed out by Nicollian and Brews [9], more than 20 properties of the MOS system can be measured and monitored using MOS capacitors.

Unfortunately, MOS capacitors could not continue to play the same role of monitor structures as the device dimensions were continuously shrunk. The need to extract the desired properties and quantities by direct measurement on MOS transistors became evident during the last two decades. This has been motivated by the appearance of several effects and phenomena inherent to MOS transistors and their progressive down scaling. Junction edge effects, hot carrier induced interface degradation and mobility degradation due to thin gate oxides are among such effects that can be only observed in MOS transistors of small gate lengths.

The MOS transistor can be used in as much the same way as an MOS capacitor to extract similar information on the properties of the $Si - SiO_2$ interface. This can be done by measuring the gate admittance of the MOS transistor with the source and drain tied together [1].

Measurement of the gate admittance of an MOS transistor has its difficulties and advantages. The main difficulties are the small gate area, parasitics due to gate-junctions overlap and two dimensional effects which could complicate data interpretation in very short channel devices. On the other hand, there are two major advantages when using an MOS transistor to conduct gate admittance measurements. The first is the very fast response of minority carriers which enable the easy interpretation of admittance measurements in weak inversion. Thus a single MOS transistor may be used to explore device and interface properties over a wide range of surface potential in the depletion and weak inversion regimes [10]. The second major advantage is that the gate current can be split into majority and minority carrier components each of which can be separately measured at the same frequency. This type of experiment has been first applied to measure the minority and majority carrier capacitances and was termed *split-capacitance* measurement [11].

Recently a generalized model for the split-current and split-admittance of the MOS transistor was developed [12] and was even used to describe large signal split-current measurements [13]. In fact, one of the most famous characterization methods, the charge pumping, is nothing but a split current measurement. A very large set of MOSFET properties can be deduced using split-admittance measurements. However, as can be expected, such measurement calls for large area devices and requires certain precautions in measurement and interpretation.

An attractive alternative to gate admittance measurement has been to make use of the MOSFET transfer admittance [14]. This implies exploiting the small signal drain current which is several orders of magnitude higher than the gate current.

In this chapter, we present small signal characterization techniques specially developed and/or adapted for MOS transistors. This is preceded by a description of the small signal equivalent circuits of MOSFETs under different bias regimes. A discussion of the minority carrier response is also presented. This is very important since it is the inversion charge frequency response which sets the limit of validity of small signal characterization methods.

---

[1]The common source/drain terminal may be left floating or may be grounded as will be seen later

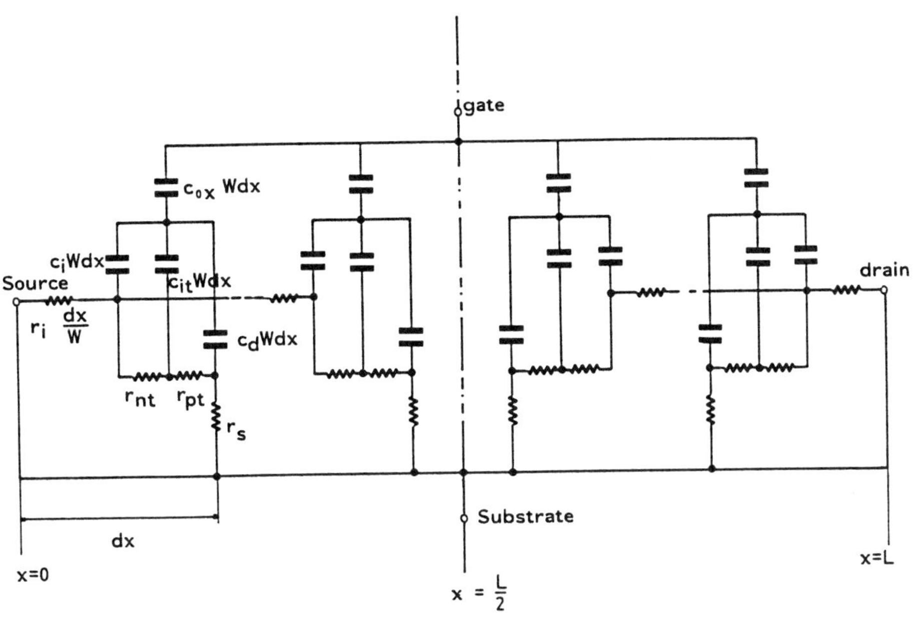

Figure 2.1: Distributed small signal equivalent circuit for an MOSFET with the source and drain terminals tied together.

2.2 Small Signal AC Model

The small signal equivalent circuit described in this section represents MOS-FETs operating under zero or very small drain bias. The effect of the drain bias on the inversion charge is therefore neglected. Under such condition, the equivalent circuit of the MOS transistor becomes nearly the same as that of an MOS capacitor.

Several workers have developed and modified distributed or transmission line models for MOS capacitors [15–18]. The purpose of using transmission line models was to analyze the lateral flow of inversion charges supplied by inverted regions outside the gate area. Similar distributed models were applied later to study the admittance of MOSFETs. The first attempts did not include the effect of interface traps in their models, whereas more recent studies did [19]. More complex distributed models including the effects of random oxide charge fluctuations and stress induced local oxide defects were presented and solved using the circuit simulator SPICE.

Figure 2.1 shows a distributed small signal equivalent circuit for an MOS-FET with grounded source, drain and substrate while the ac signal is applied to the gate. The capacitances C_{ox}, C_d, C_i and C_{it} are those associated with

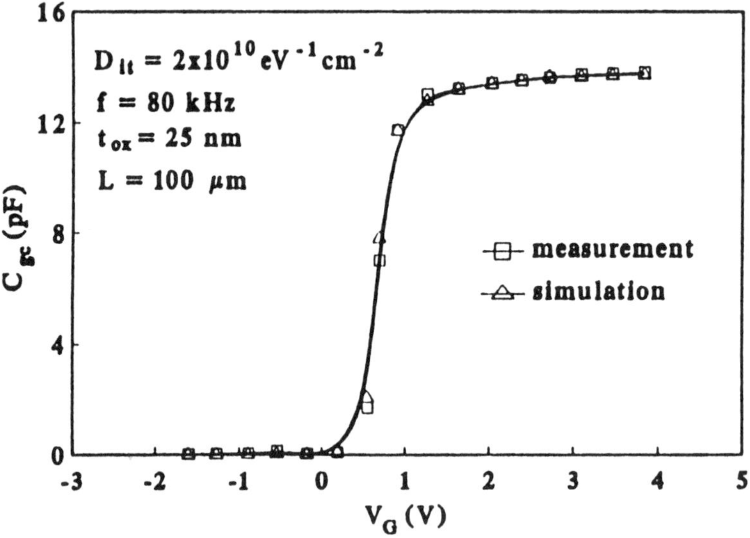

Figure 2.2: Comparison between the measured and SPICE simulated gate-channel capacitance curves for a long channel MOSFET

the gate oxide, the depletion layer, the inversion layer and the interface traps respectively. The resistance r_i is the channel sheet resistance related to the inversion charge and the effective mobility μ_{eff} by

$$r_i = \frac{1}{\mu_{eff} Q_i} \tag{2.1}$$

The resistances r_{nt} and r_{pt} are the interface trap resistances for recombination of electrons and holes whereas r_s is the substrate resistance. The equivalent circuit presented in Figure 2.1 has been successfully used as a tool to study the MOSFET gate admittance and the channel frequency response of MOS transistors. Chow and Wang [19] have provided an analytical resolution of the above model after neglecting the interface trap conductance. They have then used their model to study the effects of high field stressing on the channel frequency response of MOSFETs. The complete transmission line model can be numerically solved using SPICE. Such an approach provides an accurate representation of the MOSFET under ac operation. Figures 2.2 and 2.3 show a comparison between the measured gate-channel capacitance and conductance and the ones simulated using SPICE [20]. As mentioned earlier, numerical simulation allows the resolution of the complete transmission line model while

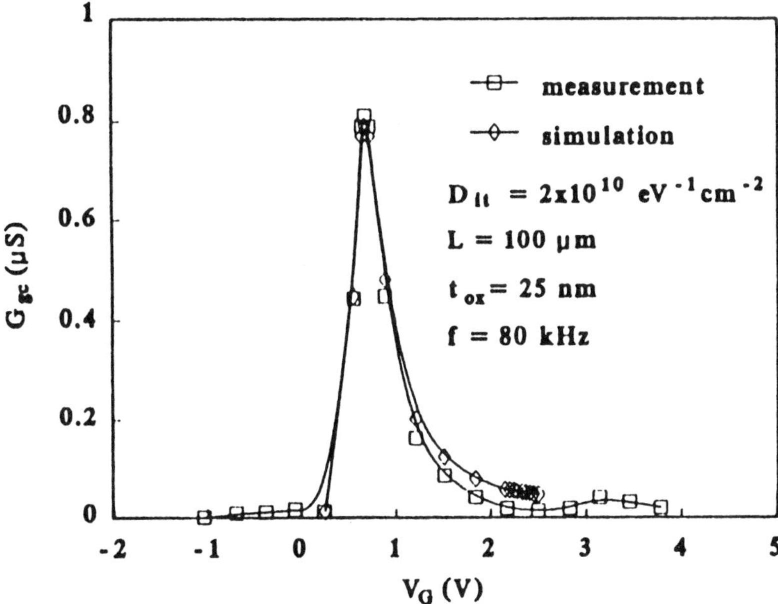

Figure 2.3: Comparison between the measured and SPICE simulated gate-channel conductance curves for a long channel MOSFET

adding more complex features such as random distributions of oxide charges and stress induced local interface traps.

Although distributed models are important for modeling the channel frequency response, they are quite inadequate for the extraction of interface and device parameters. Indeed, practical characterization methods are all based on more simple lumped ac models.

Figure 2.4 shows a lumped ac small signal model with two symmetrical parts ; one connected to the source and the other to the drain. For uniform distributions of inversion and interface trap charges along the channel, the two symmetrical parts can be gathered in a single part as shown in Figure 2.5.

2.3 Channel Frequency Response of MOSFETs

The frequency response of an MOS transistor is characterized in the literature by several parameters, in a way that could be somewhat confusing. Terms like "gate-channel time constant", "inversion layer time constant" and "intrinsic cutoff frequency" are alternately used to describe the ac small signal frequency response of MOSFETs. It is our aim in this section to review the definitions of these parameters and the information provided by each one of them.

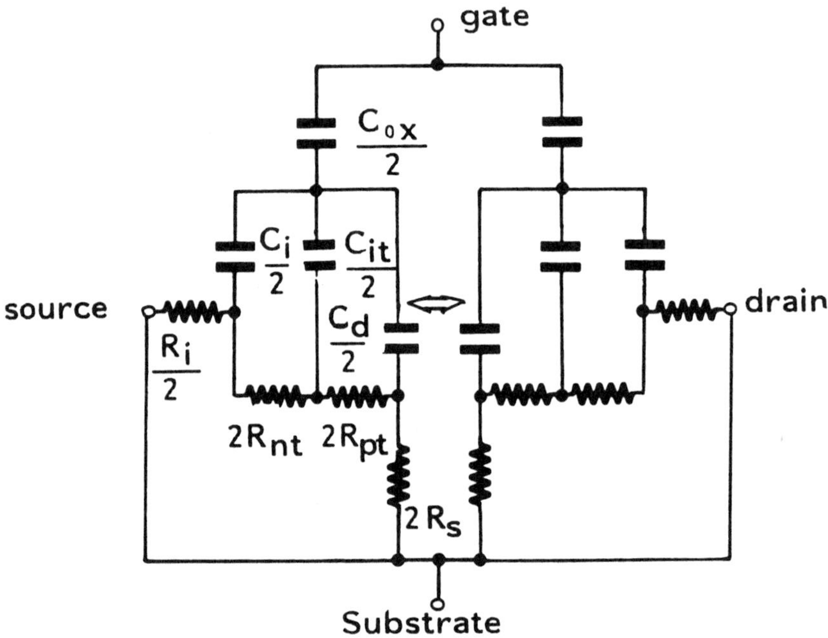

Figure 2.4: Simplified small signal equivalent circuit for an MOSFET of a relatively short channel.

All the above three parameters predict, more or less, comparable limits to the high frequency response of the device. However, the inversion layer time constant as well as the gate-channel time constant are only related to the input gate admittance of the device. Both parameters are defined for operation at very small drain bias. On the other hand, the intrinsic cutoff frequency is usually defined for operation in the saturation regime where the device is normally used as an amplifier.

The inversion layer time constant is defined as

$$\tau_i = R_i C_i = \frac{L^2 C_i}{4\mu_{eff} Q_i} \tag{2.2}$$

The inversion capacitance C_i is given by the following expressions

$$C_i \approx \begin{cases} \frac{Q_i}{(2kT/q)} & \text{strong inversion} \\[2ex] \frac{Q_i}{(kT/q)} & \text{weak inversion} \end{cases} \tag{2.3}$$

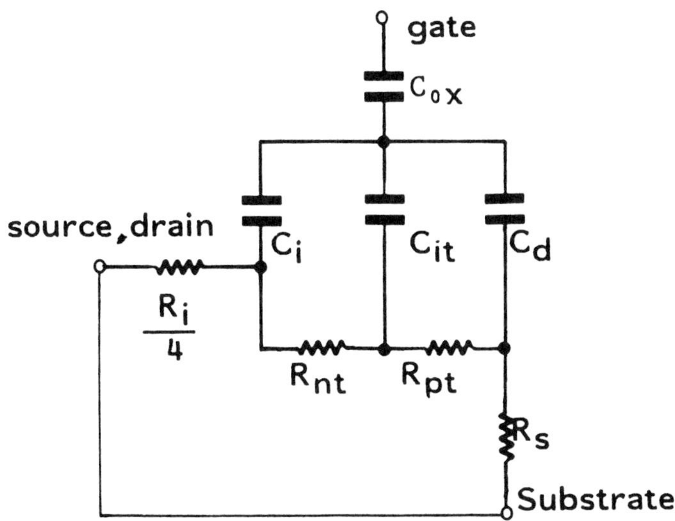

Figure 2.5: Lumped small signal equivalent circuit for uniform distributions of the inversion and interface trap charges.

Consequently, τ_i becomes

$$
\tau_i = \begin{cases} \dfrac{L^2}{4\mu_{eff}(kT/q)} & \text{weak inversion} \\[3ex] \dfrac{L^2}{4\mu_{eff}(2kT/q)} & \text{strong inversion} \end{cases}
\tag{2.4}
$$

Notice that τ_i is gate voltage independent. As can be deduced from its definition, the inversion time constant is a measure of the speed of the formation of the intrinsic inversion channel and not the speed of the device as a whole.

Unlike τ_i, the gate channel time constant τ_{gc} can be conveniently determined from experimental measurements as it characterizes the transistor's input admittance which is a directly measurable quantity. This time constant is given by

$$
\tau_{gc} = \begin{cases} \dfrac{L^2}{4\mu_{eff}(kT/q)} & \text{weak inversion} \\[3ex] \dfrac{L^2}{4\mu_{eff}(V_g-V_t)} & \text{strong inversion} \end{cases}
\tag{2.5}
$$

In weak inversion, both τ_{gc} and τ_i are given by the same expression; this is because the input capacitance is approximately equal to C_i in weak inversion. However, in strong inversion, the input capacitance is nothing but C_{ox} and τ_{gc} becomes inversely proportional to $(V_g - V_t)$.

The above expressions for τ_i and τ_{gc} are deduced in absence of interface traps. In strong inversion, the presence of interface traps will not lead to a significant difference unless their density is extremely high. However, in weak inversion, the effect of interface traps cannot be neglected. In this case, the gate-channel admittance becomes more complex and cannot be represented by a single time constant. It is the objective of the next section to analyze the gate-channel admittance in presence of interface traps for the purpose of characterizing the $Si - SiO_2$ interface.

Finally, the intrinsic cutoff frequency f_T is defined as the frequency at which the short circuit current gain of the transistor equals unity. The cutoff frequency f_T is expressed as

$$f_T = \frac{g_m}{2\pi C_g} \simeq \frac{1}{2\pi} \frac{3\mu_{eff}(V_g - V_t)}{2(1 + \delta)L^2} \tag{2.6}$$

where C_g is the total gate capacitance and δ is a function of the body effect coefficient [21].

2.4 The Split Admittance Technique

2.4.1 Concept of Split-Current

When a time varying signal is applied to the gate of an MOS transistor connected as shown in Figure 2.6, the variation of the surface potential calls for changes in the magnitudes of the different semiconductor charges.

A very interesting feature of the MOS transistor is that variations in the majority and minority carrier charges are separated and collected through different device terminals. Majority carriers are supplied by the substrate while minority carriers are provided by the source and drain regions. In this way the transistor acts as a splitter which divides the input gate current into minority and majority carrier components. This phenomenon of current splitting is at the heart of the charge pumping technique [22, 23] and can be equally exploited in small signal measurements for detailed characterization of MOSFET properties.

Unlike MOS capacitors, minority carriers can follow rapid gate signal variations in MOS transistors. This is because minority carriers are now provided by the highly doped source and drain regions. This makes small signal characterization of interface traps feasible not only in depletion but also in weak inversion where the traps can easily interact with the minority carrier band [10].

2.4.2 Split-Admittance Model

Let us consider in our analysis the case of an n-channel MOSFET where a small harmonic signal is applied to the gate while the source, drain and substrate terminals are connected to ground (switch S connected to H in Figure 2.6). The gate voltage at any time instant can be expressed as

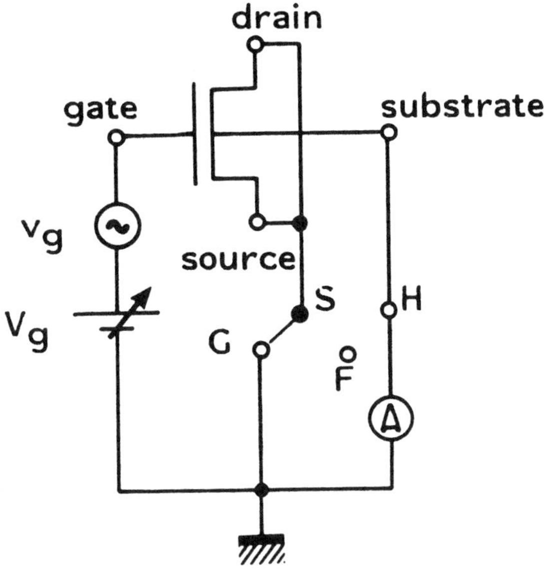

Figure 2.6: Schematic diagram of the split-current measuring circuit.

$$V_g(t) = \Phi_{ms} + \psi_s - \frac{1}{C_{ox}}(Q_i + Q_b + Q_{ox} + Q_{it}) \tag{2.7}$$

where Q_i, Q_b, Q_{ox} and Q_{it} are the inversion, bulk, fixed oxide and interface trap charge densities respectively. Also the charge neutrality in an MOS system relates the gate charge Q_g to other charges by

$$Q_g(t) + Q_b(t) + Q_i(t) + Q_{it}(t) + Q_{ox}(t) = 0 \tag{2.8}$$

Differentiating Equations 2.7 and 2.8 with respect to time yields

$$\frac{dV_g}{dt} = \frac{1}{C_{ox}}[(C_b + C_i + C_{ox})\frac{d\psi_s}{dt} + (J_n + J_p)] \tag{2.9}$$

$$\frac{dQ_g}{dt} - (C_b + C_i)\frac{d\psi_s}{dt} - (J_n + J_p) = 0 \tag{2.10}$$

where

$$C_b = -\frac{dQ_b}{d\psi_s} \tag{2.11}$$

$$C_i = -\frac{dQ_i}{d\psi_s} \tag{2.12}$$

$$J_n + J_p = -\frac{dQ_{it}}{dt} \tag{2.13}$$

The quantity dQ_{it}/dt is the net interface trap recombination current density while J_n and J_p are its electron and hole components respectively. The electron current J_n results from the interaction of traps situated in the upper half of the energy gap with the conduction band while J_p arises from the communication of traps situated in the lower half with the valence band. Consequently, the interface trap recombination current will be dominated by J_n when the gate is biased in inversion and by J_p for gate voltages in depletion or accumulation. Making use of the above equations, one can deduce the gate current density J_g as well as the split-current densities J_{sub} and $J_{s,d}$ measured in the substrate and the source/drain terminals respectively.

$$J_g = \frac{dQ_g}{dt} = C_g \frac{dV_g}{dt} + \frac{C_{ox}}{C_{ox} + C_b + C_i}(J_n + J_p) \tag{2.14}$$

$$J_{sub} = C_b \frac{d\psi_s}{dt} + J_p = C_{gb}^0 \frac{dV_g}{dt} + \left(1 - \frac{C_{gb}^0}{C_{ox}}\right) J_p - \frac{C_{gb}^0}{C_{ox}} J_n \tag{2.15}$$

$$J_{s,d} = C_i \frac{d\psi_s}{dt} + J_n = C_{gc}^0 \frac{dV_g}{dt} + \left(1 - \frac{C_{gc}^0}{C_{ox}}\right) J_n - \frac{C_{gc}^0}{C_{ox}} J_p \tag{2.16}$$

where C_{gc}^0 and C_{gb}^0 are the gate-channel and gate-bulk capacitances respectively, in the absence of interface traps[2],

$$C_{gc}^0 = \frac{C_i C_{ox}}{C_{ox} + C_b + C_i} \quad ; \quad C_{gb}^0 = \frac{C_b C_{ox}}{C_{ox} + C_b + C_i} \tag{2.17}$$

In Equations 2.14, 2.15 and 2.16, the first term on the right hand side is the displacement current density while the other terms arise from the contributions of the electron and hole recombination currents.

Now we are going to deduce the split admittance components Y_{gc} and Y_{gb} in both the depletion and weak inversion regimes. The gate-channel admittance Y_{gc} and the gate-bulk admittance Y_{gb} are defined as

$$Y_{gc} = \frac{i_{s,d}}{v_g}\Big|_{v_b=0} \quad ; \quad Y_{gb} = \frac{i_{sub}}{v_g}\Big|_{v_{s,d}=0} \tag{2.18}$$

Depletion

If a small harmonic signal δV_g is applied to the gate of an MOSFET while the device is being biased in depletion, then the small signal substrate and source/drain currents can be derived using Equations 2.15 and 2.16.

The source/drain current is obviously zero in this case since the channel is absent and as a result Y_{gc} vanishes in depletion. On the other hand, the small signal substrate current i_{sub} is expressed as

$$i_{sub} = C_b \frac{d\psi_s}{dV_g}(\jmath\omega)\delta V_g + \frac{Y_{pt}}{\jmath\omega}\frac{d\psi_s}{dV_g}(\jmath\omega)\delta V_g \tag{2.19}$$

---

[2] Generalized expressions for these capacitances will be developed later

where Y_{pt} is the equivalent interface trap admittance in depletion [8],

$$Y_{pt} = G_{pp} + \jmath\omega C_{pp} \tag{2.20}$$

There are several models for the interface trap admittance Y_{pt} [9]. If a continuous energy distribution of interface traps is considered then

$$Y_{pt} = \frac{qD_{it}}{2\tau}\ln(1 + \omega^2\tau^2) + \jmath\frac{qD_{it}}{\tau}\arctan(\omega\tau) \tag{2.21}$$

The subscript pt indicates that such interface trap admittance arises from the interaction of traps with the valence band by means of hole emission and capture mechanisms. The interface trap density D_{it} is a function of energy whereas the time constant τ is a function of the trap capture cross section σ, the thermal velocity of carriers v_{th} and the carrier concentration at the $Si - SiO_2$ interface.

Making use of Equation 2.7, the rate of change of the surface potential with respect to the gate voltage is given by

$$\frac{d\psi_s}{dV_g} = \frac{C_{ox}}{C_b + C_{ox} + C_{pp} + G_{pp}/\jmath\omega} \tag{2.22}$$

substituting $d\psi_s/dV_g$ in Equation 2.19, yields the following expression for the gate-bulk admittance

$$Y_{gb} = [G_{pp} + \jmath\omega(C_b + C_{pp})]\frac{C_{ox}}{C_b + C_{ox} + C_{pp} + G_{pp}/\jmath\omega} \tag{2.23}$$

For most practical cases, the interface trap conductance G_{pp} is much less than the total reactance $\omega(C_{ox} + C_b + C_{pp})$ especially for VLSI devices where C_{ox} is relatively high. As a result, Y_{gb} can be simplified to

$$\begin{aligned} Y_{gb} &= G_{gb} + j\omega C_{gb} \tag{2.24} \\ &= \frac{G_{pp}C_{ox}^2}{(C_{ox} + C_b + C_{pp})^2} + \jmath\omega\frac{C_{ox}(C_b + C_{pp})}{C_{ox} + C_b + C_{pp}} \tag{2.25} \end{aligned}$$

where G_{gb} and C_{gb} are termed the gate-bulk conductance and capacitance respectively.

Weak inversion

In weak inversion, both the gate-bulk as well as the gate channel admittances have finite values. The reason that the gate-bulk admittance does not vanish in weak inversion is the existence of a non-zero depletion capacitance which ensures the coupling between the substrate and other device terminals. The gate-bulk and gate-channel admittances can be deduced from Equations 2.15

46

and 2.16 after putting $J_p = 0$ because the interface traps will interact only with the conduction band in weak inversion by the emission and capture of electrons.

Therefore, following a similar procedure as in the depletion regime, the weak inversion Y_{gb} and Y_{gc} can expressed as

$$Y_{gb} = -\frac{G_{pn}C_bC_{ox}}{(C_{ox} + C_i + C_b + C_{pn})^2} + \jmath\omega\frac{C_bC_{ox}}{C_{ox} + C_i + C_b + C_{pn}} \tag{2.26}$$

$$Y_{gc} = \frac{G_{pn}C_{ox}(C_{ox} + C_b)}{(C_{ox} + C_i + C_b + C_{pn})^2} + \jmath\omega\frac{C_{ox}(C_i + C_{pn})}{C_{ox} + C_i + C_b + C_{pn}} \tag{2.27}$$

where

$$Y_{nt} = G_{pn} + \jmath\omega C_{pn} \tag{2.28}$$

is the equivalent parallel interface trap admittance in weak inversion. As in depletion, the interface trap conductance (here G_{pn}) has been neglected against the total reactance $\omega(C_{ox} + C_i + C_b + C_{pn})$.

2.4.3 Admittance-Voltage Characteristics

Split-Capacitance

Figure 2.7 shows a pair of split C-V curves measured at a small signal frequency of 1 MHz for a transistor of channel length L $= 3\,\mu$m. The gate-channel capacitance C_{gc} is offset by a value Δ above zero in accumulation. This corresponds to the gate/junctions overlap capacitance. In strong inversion, C_{gc} saturates at a value equal $C_{ox} + \Delta$.

As for C_{gb}, it approaches zero in strong inversion unlike the gate-bulk capacitance of a simple MOS capacitor which exhibits a constant minimum value in strong inversion. This behavior is due to the presence of the inversion capacitance C_i which abruptly pulls C_{gb} towards zero at the onset of strong inversion.

If however the source and drain terminals were left floating and C_{gb} measured at a relatively high frequency, the measured capacitance will be nothing but the conventional high frequency C-V known for MOS structures. This is shown in Figure 2.8 which depicts the gate-bulk capacitance with the source and drain grounded and that while they are floating, both measured at 1 MHz. It is clear that curve b (source and drain open) can be used to extract parameters such as the flat-band voltage and substrate doping concentration in the same way as with MOS diodes.

Split-Conductance

The existence of a finite depletion capacitance in weak inversion allows current to flow through the substrate. Therefore, a finite gate-bulk conductance G_{gb} exists in weak inversion as well as in depletion whereas the gate-channel conductance only exists in weak inversion due to the absence of coupling between the

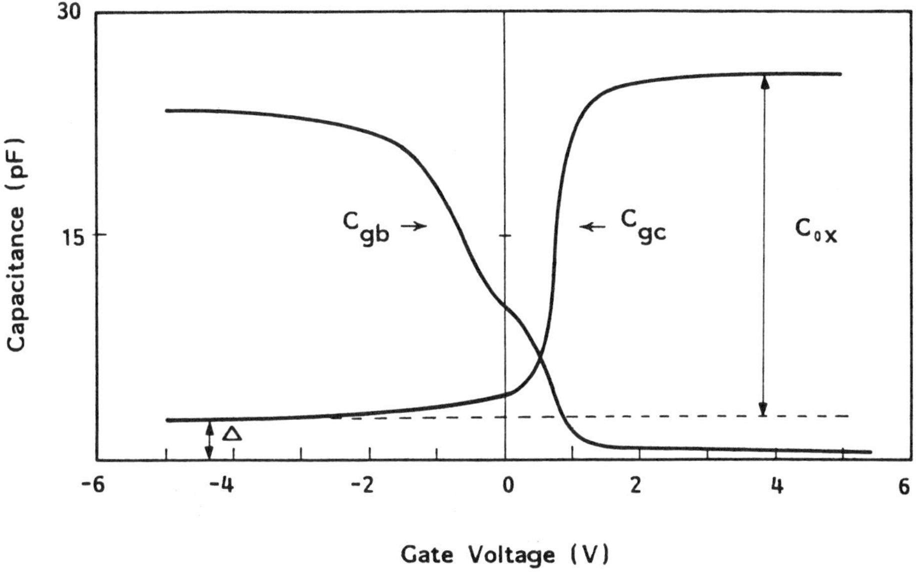

Figure 2.7: Split-capacitance curves for an MOSFET having a channel length of 3 μm.

source/drain terminals and the substrate in depletion. Figure 2.9 shows typical split conductance curves measured at a small signal frequency of 50 kHz. Notice that G_{gb} passes through two peaks; the first one is situated in depletion and is classically attributed to the interaction of interface traps with the valence band as in MOS capacitors. The second peak however, is negative in magnitude and coincides with the positive gate-channel conductance peak in weak inversion. This negative gate-bulk conductance is attributed to current flowing from the source/drain terminals into the substrate. Notice that the gate-channel conductance represents the delay associated with the interaction of the interface traps with the conduction band. Such conductance peak can only be observed in MOS transistors where minority carriers are rapidly supplied by the source and drain junctions.

Gate Admittance

The total gate admittance is the sum of the gate-channel and gate-bulk admittances

$$Y_m = Y_{gc} + Y_{gb} \tag{2.29}$$

The gate capacitance and conductance characteristic curves are shown in

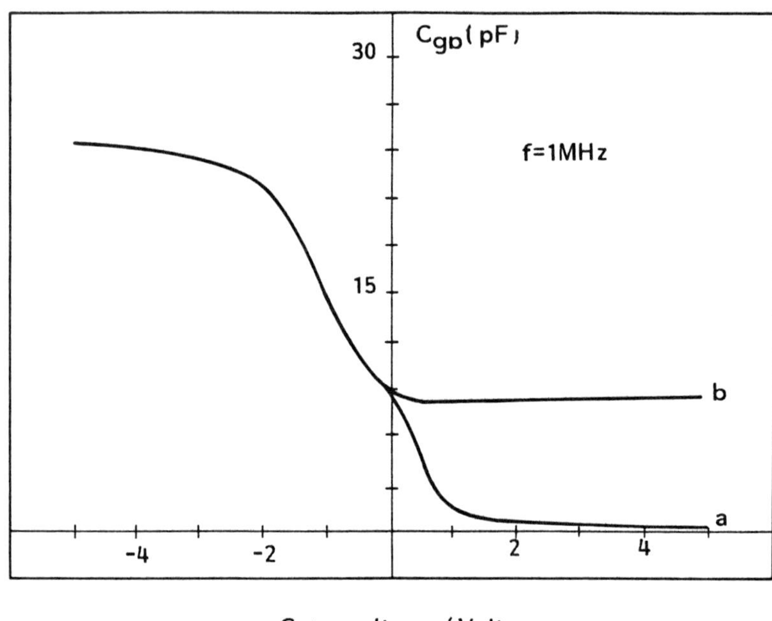

Gate voltage (Volt)

Figure 2.8: Variation of the high frequency gate-bulk capacitance with gate bias with the source and drain grounded (curve a) and floating (curve b).

Figure 2.10 for two different frequencies. Notice that the two conductance peaks shift towards higher values of gate bias as the frequency increases. This implies that, at high frequency, carriers can only interact with trap levels closer to the band edges because of the fast response of these traps. On the other hand, at very low frequencies, the carrier bands will be able to communicate with trap levels in the mid-gap region and the two conductance peaks will tend to merge in a single peak. Analysis of such low frequency conductance curves is therefore relatively complex and may even become more complicated due to bulk trap loss which could be dominant over that of interface traps in some cases.

Inverse Capacitance and Conductance Relations

The interest of the split admittance model resides in the fact that it provides an extremely useful set of relations that can be employed to conduct an extensive characterization of MOSFET properties.

Starting from the expressions which describe the split admittance components in weak inversion and solving for the intrinsic MOSFET capacitances and

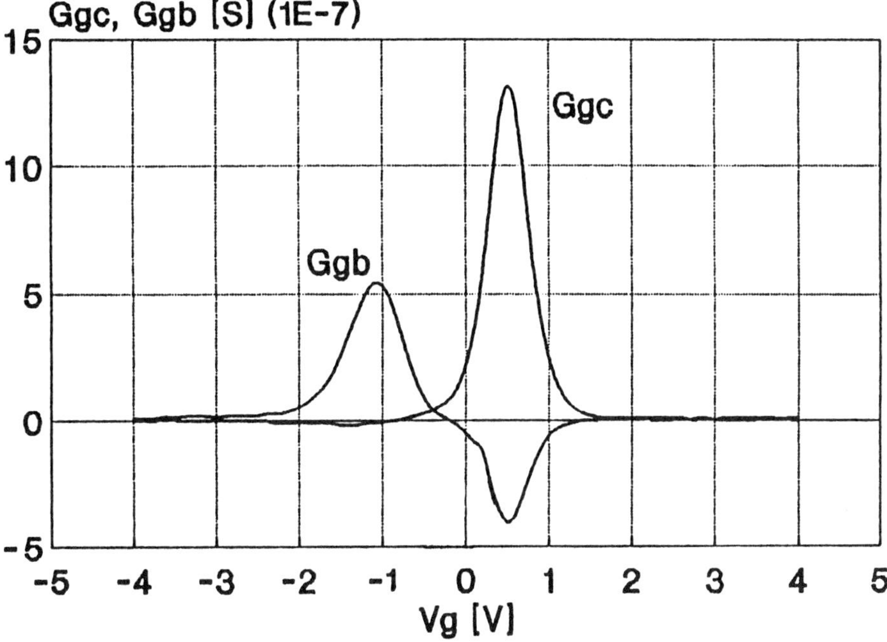

Figure 2.9: Split-conductance curves for an MOSFET of $L = 3\ \mu m$ measured at $f = 50\ kHz$.

conductances, one obtains the following set of equations

$$C_{ox} = C_{gc} - \frac{C_{gb}}{G_{gc}/G_{gb}} \tag{2.30}$$

$$C_b = -\frac{C_{ox}G_{gb}}{G_{gb} + G_{gc}} \tag{2.31}$$

$$\frac{G_{pn}}{\omega} = \frac{\omega C_{ox}^2(G_{gc} + G_{gb})}{(G_{gb} + G_{gc})^2 + \omega^2[C_{ox} - (C_{gc} + C_{gb})]^2} \tag{2.32}$$

$$\frac{C_{pn} + C_b + C_i}{C_{ox}} = \frac{\omega^2(C_{gc}+C_{gb})[C_{ox} - (C_{gc}+C_{gb})] - (G_{gb} + G_{gc})^2}{(G_{gb}+G_{gc})^2 + \omega^2[C_{ox} - (C_{gc} + C_{gb})]^2} \tag{2.33}$$

It is obvious that the above set of equations can be used to extract C_{ox}, C_b, G_{pn}/ω and the sum $(C_{pn} + C_i)$ from the results of split admittance measurements. Moreover, since C_{pn} is related to G_{pn} through different interface trap models [9], both C_{pn} and C_i can be separately evaluated at any frequency and any gate bias.

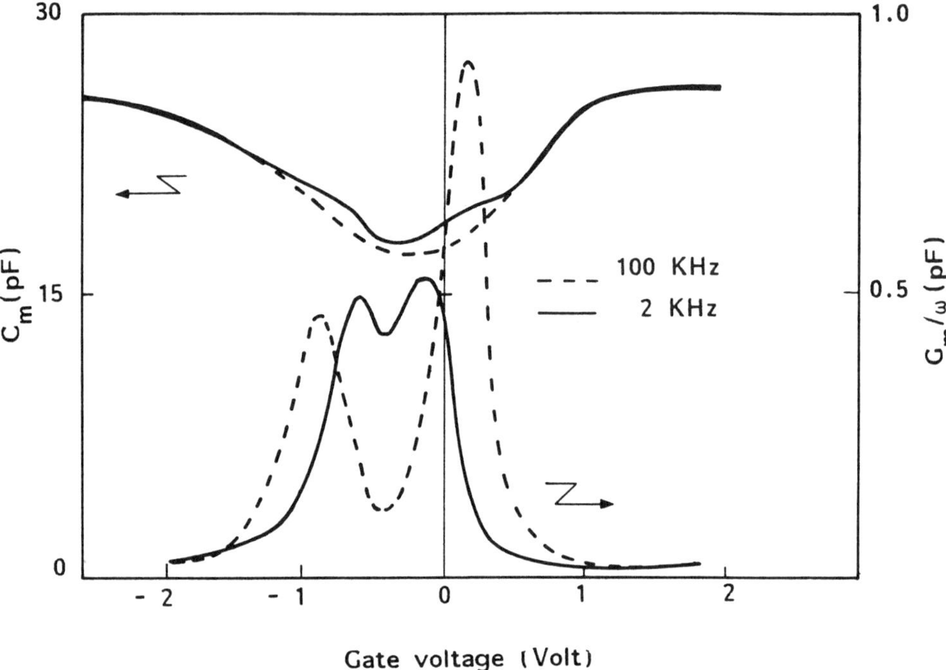

Figure 2.10: Gate capacitance and conductance curves measured at two different frequencies.

Using the same approach in the depletion regime, the following two equations can be easily deduced

$$\frac{G_{pp}}{\omega} = \frac{\omega C_{ox}^2 G_{gb}}{G_{gb}^2 + \omega^2 (C_{ox} - C_{gb})^2} \tag{2.34}$$

$$\frac{C_{pp} + C_b}{C_{ox}} = \frac{\omega^2 C_{gb}(C_{ox} - C_{gb}) - G_{gb}^2}{G_{gb}^2 + \omega^2 (C_{ox} - C_{gb})^2} \tag{2.35}$$

$$\tag{2.36}$$

2.4.4 Applications of Split-Admittance Measurements

Interface Trap Characterization

Full characterization of interface traps over a wide range of the Si energy gap can be achieved by direct split admittance measurement on a single MOSFET. The split admittance model extends the application of the classical conductance technique [8] to the weak inversion regime in addition to depletion. As a result, the energy distributions of the interface trap density, time constants and capture

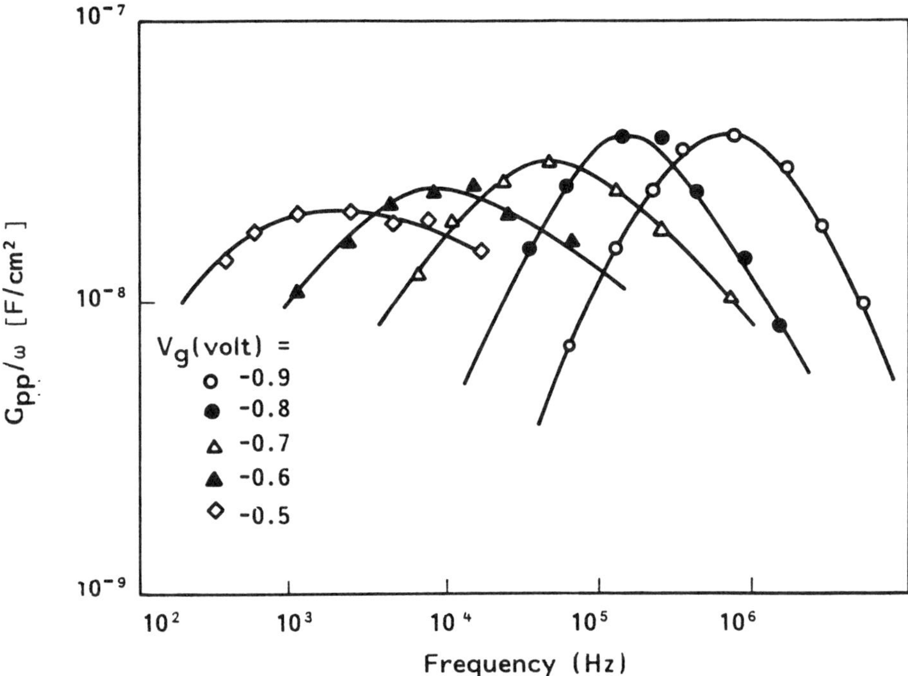

Figure 2.11: Experimental G_{pp}/ω versus frequency for various gate voltages in depletion.

cross sections can be extracted from G_{pn}/ω and G_{pp}/ω curves using the interface trap admittance model for a continuum of trap levels with or without random potential fluctuations.

Figures 2.11 and 2.12 show the conductance curves versus frequency with the gate bias as a parameter in depletion and weak inversion. Based on the interface trap admittance model for a continuum of trap levels, both the density D_{it} and the time constant τ_n (or τ_p) of interface traps are determined from the magnitudes and positions of the maxima of these curves. Alternatively, the measured conductance curves can be fitted with the statistical model of the interface trap admittance [8] in order to account for potential fluctuations induced by random oxide charge distributions. This fitting provides : (a) the standard deviation of the potential distribution, (b) the interface trap density and (c) the interface trap time constant τ_n or τ_p, all at a given gate bias. The energy distribution of the interface trap density and time constant can be constructed after determining the surface potential as a function of gate voltage [1, 3]. The traps capture cross sections for electrons σ_n and holes σ_p can also be obtained using the following relations

$$\tau_n = \frac{1}{v_{th}\sigma_n n_s} \tag{2.37}$$

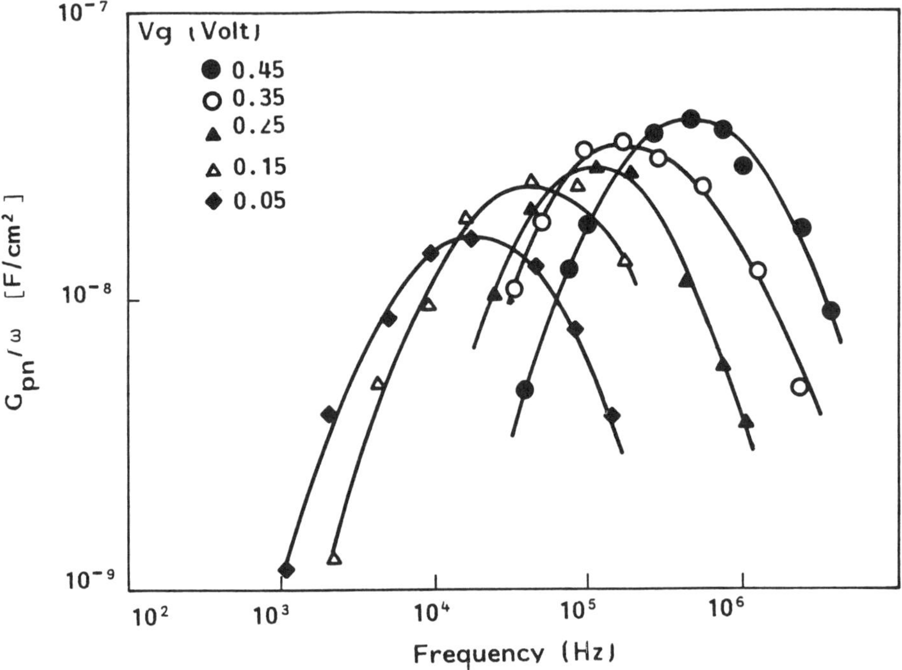

Figure 2.12: Experimental G_{pn}/ω versus frequency for various gate voltages in weak inversion.

$$\tau_p = \frac{1}{v_{th}\sigma_p p_s} \tag{2.38}$$

with v_{th} being the thermal velocity and n_s and p_s the electron and hole surface concentrations respectively. Examples of the energy profiles of the interface trap parameters are shown in Figures 2.13 and 2.14.

Characterization of MOSFET Channel Mobility

Transverse and lateral field dependencies of the carrier mobility in MOSFETs are of great importance specially when such devices are scaled down to very small dimensions in VLSI circuits. Several studies have been proposed to characterize the effective channel mobility as a function of the effective transverse field intensity E_{eff} [24–27]. The most widely used method for the determination of the MOSFET channel mobility is based on the evaluation of the inversion charge as a function of gate bias by measuring the gate-channel capacitance C_{gc} and then integrating it as a function of gate bias

$$Q_i(V_g) = \int_{-\infty}^{V_g} C_{gc}(V_g)\,dV_g \tag{2.39}$$

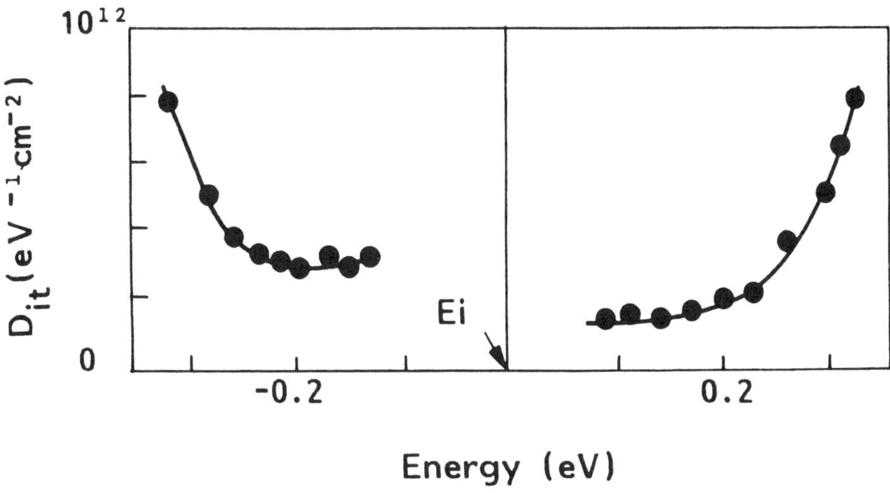

Figure 2.13: Interface trap density profile versus energy in the Si band gap as obtained using the split admittance technique.

The obtained inversion charge density is then substituted into the drain current expression under small drain bias

$$I_d = \mu_{eff} Q_i \frac{W}{L} V_d \tag{2.40}$$

and the effective mobility is obtained for each gate bias. The corresponding effective transverse field can be deduced once the inversion and depletion charge densities are known using the following expression [25],

$$E_{eff} = \frac{Q_i/2 + Q_b}{\epsilon_{si}} \tag{2.41}$$

It is obvious that this method is not sufficiently accurate in case of relatively high interface trap densities where the inversion charge obtained from the integration of C_{gc} will be overestimated due to the contribution of the interface trap charge. This will lead to errors in the evaluation of μ_{eff} as well as E_{eff} especially in weak inversion where Q_{it} becomes comparable to Q_i. Moreover, the integration errors become very high as the threshold voltage is approached due to the steep descent of C_{gc}.

One way to eliminate the effect of interface traps is to measure C_{gc} at a relatively high frequency. This method is only applicable in strong inversion where the channel formation delay is very small. In some cases, however, it is interesting to study the effective mobility in weak inversion and in the vicinity

54

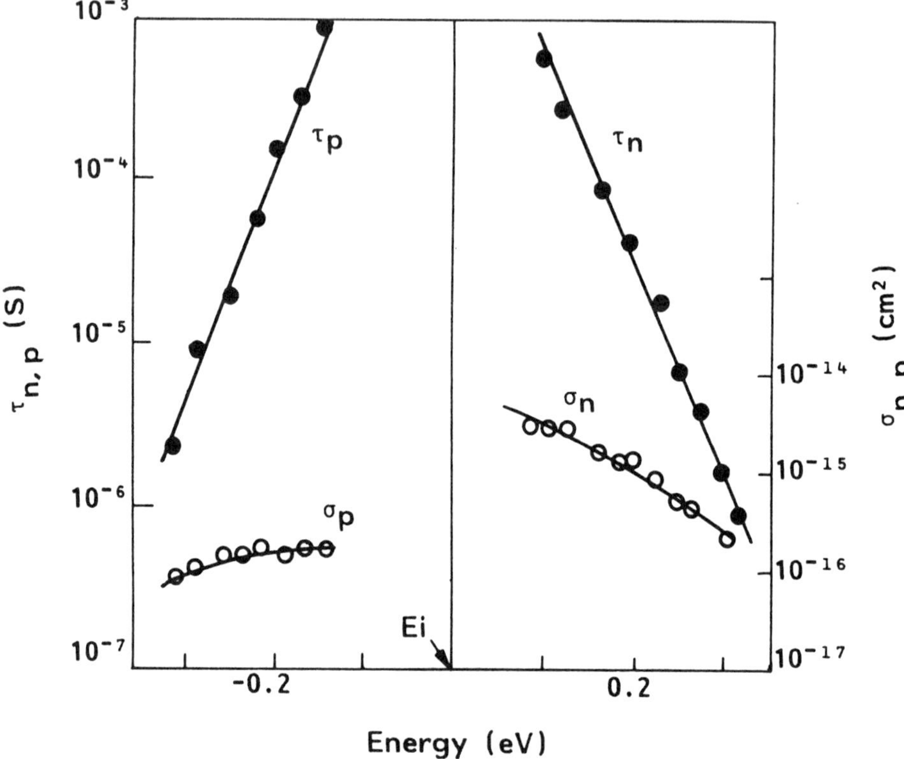

Figure 2.14: Interface trap time constants $\tau_{p,n}$ (filled circles) and capture cross sections (empty circles) versus energy in the Si band gap.

of the threshold voltage to evaluate the effects of oxide charge and channel nonuniformities.

As pointed out earlier, the split admittance technique enables the separation of the inversion capacitance C_i from that of the interface traps C_{pn} in weak inversion. Consequently, a purified gate channel capacitance C'_{gc} can be constructed excluding the interface trap charge contribution

$$C'_{gc} = C^0_{gc} = \frac{C_i C_{ox}}{C_{ox} + C_b + C_i} \tag{2.42}$$

Now, the inversion layer charge Q_i can be accurately evaluated as

$$Q_i(V_g) = \int_{-\infty}^{V_g} C'_{gc}(V_g)\, dV_g \tag{2.43}$$

The effective channel mobility can then be calculated using the drain current expression (Equation 2.40).

Depth Profiling of Substrate Doping Concentration

The C-V plot of an MOS capacitor has been frequently used to probe the doping concentration profiles of semiconductor substrates. For this purpose, the device is biased in depletion and the resulting capacitance is the series combination of C_{ox} and C_b. This capacitance is differentiated with respect to the gate voltage to yield

$$\frac{d}{dV_g}(C/C_{ox}) = -\left(\frac{C}{C_{ox}}\right)^3 \frac{\epsilon_{ox}^2 \epsilon_0}{\epsilon_{si} q N_A t_{ox}^2} \qquad (2.44)$$

At any point on the C-V plot, the obtained doping concentration is that at the edge of the depletion layer established by the applied bias. The corresponding depletion layer width can be calculated once N_A is known.

This kind of measurement can be directly applied to an MOS transistor by measuring the gate bulk capacitance C_{gb} with the source and drain floating in order to eliminate the effect of the inversion capacitance. Furthermore, this capacitance should be measured at high frequency to exclude the interface trap capacitance. The obtained high frequency C-V curve can be used to extract the same information originally deduced using MOS capacitors.

Before concluding this part on split admittance measurements, it is important to point out that such kind of measurements necessitates relatively large area MOSFETs. For devices having channel lengths in the range of 1 micron, the channel width has to be in the order of 100 microns or more so as to be able to conduct capacitance or conductance measurements. However, the plenitude of information offered by the split admittance technique encourages the use of large width MOSFETs as monitor structures. Such devices can be used to study interface defects induced by hot-carrier stress, mobility degradation in thin gate oxides and many other phenomena associated with short channel MOSFETs by direct measurements on a single transistor.

2.5   Dynamic Transconductance

Since the early seventies, several workers have tried to extract information about the interface properties of MOSFETs from dynamic transconductance measurements [28–31]. The idea seemed attractive since the transconductance, unlike gate admittance, can be directly measured on small area devices. In fact, the sensitivity of dynamic transconductance measurements increases as the channel length decreases. However, the lack of a simple analytical model relating the transfer admittance to the interface trap parameters hindered any progress in the development of a reliable characterization technique.

It has been until the last decade when a simple model providing a direct correlation between the MOSFET dynamic transcondutance $g_m(\omega)$ and the interface trap admittance was introduced [32]. This model was then used to develop an interface trap characterization technique based on measuring the imaginary part of the transfer impedance (*i.e.* $\Im\{1/g_m(\omega)\}$). The method

56

was found to be quite sensitive and has been capable of providing the same information as the conductance technique in weak inversion [33].

Later on, this technique was adapted to depletion mode devices [34] and was used by several authors to characterize SOI MOSFETs. In 1991, a new variant of dynamic transconductance, based on measuring the real part of $g_m(\omega)$, was proposed [35]. In what follows we will present the theory of the dynamic transconductance in enhancement and depletion MOSFETs. Then we will introduce the different variants of dynamic transconductance measuring techniques and discuss their advantages and drawbacks.

2.5.1   Enhancement MOSFET Dynamic Transconductance

Conventional capacitance and conductance techniques are based on measuring the gate admittance or the split admittance of MOSFETs as shown earlier. In other words, the small signal gate current i_c is measured in presence of a small signal gate voltage. Such current is evidently very small specially in small area devices where the ac gate current can be in the order of a few pico-amperes.

In contrast to that, the dynamic transconductance involves measuring the small signal drain current i_d which is several orders of magnitude higher than i_c. It is straightforward to prove that

$$g_m = \left(\frac{i_d}{i_c}\right) Y_m \tag{2.45}$$

where $Y_m = (G_m + j\omega C_m)$ is the measured gate admittance. The above relation indicates that g_m is directly related to Y_m through the amplification factor i_d/i_c and can therefore be used to extract the same information offered by Y_m.

We will consider in the following analysis an n-channel MOSFET operating in weak inversion. However, the final results can be readily applied to p-channel devices.

In case of small drain bias ($V_d \ll kT/q$), the transfer admittance $g_m(\omega)$ is obtained by differentiating Equation 2.40 with respect to V_g

$$g_m(\omega) = \left(\frac{W}{L}\right) V_d \left[\mu_{eff}\frac{dQ_i}{d\psi_s} + Q_i\frac{d\mu_{eff}}{d\psi_s}\right]\frac{d\psi_s}{dV_g} \tag{2.46}$$

Making use of the instantaneous charge conservation equation in an MOS system and following the same approach as in [32], the dynamic transconductance can be written as

$$g_m(\omega) = \left(\frac{I_d}{kT/q}\right)\frac{C_{ox}}{C_{ox} + C_i + C_b + Y_{nt}/j\omega} \tag{2.47}$$

Note that the above model has been reached after assuming a constant mobility in weak inversion. Now, we are going to present two different alternatives for interface trap characterization. The first of these methods is based on the measurement of the imaginary part of g_m^{-1} and is called the G_p/ω method. The second technique is based on measuring the real part of g_m and is known as the high/low frequency transconductance method.

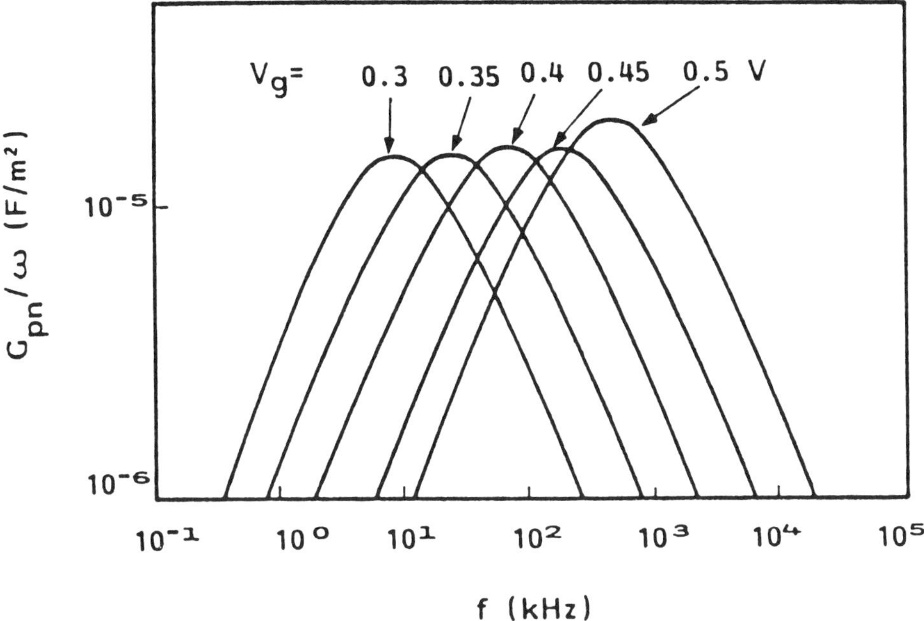

Figure 2.15: Typical curves for G_p/ω versus frequency as obtained from the imaginary part of g_m^{-1}.

The G_p/ω Method

It is obvious from the dynamic transconductance model that the imaginary part of $g_m^{-1}(\omega)$ provides a direct access to the interface trap conductance. In fact, the famous bell-shaped frequency dispersion of G_p/ω is readily obtained by measuring $\Im\{g_m^{-1}(\omega)\}$ and the static drain current I_d,

$$\frac{G_p}{\omega} = \frac{qC_{ox}I_d}{kT}\,\Im\{g_m^{-1}(\omega)\} \tag{2.48}$$

Figure 2.15 shows the measured variation of the imaginary part of $g_m^{-1}(\omega)$ normalized by the factor $kT/(qC_{ox}I_d)$ for several values of gate bias in weak inversion. The curves have the same shape and behavior as those usually obtained using the conductance technique [9] on MOS capacitors. However, dynamic transconductance measurements are much easier than MOS gate admittance measurements. Moreover, the G_p/ω curves are readily obtained from the measurement whereas in case of the conductance technique several manipulations and compensation or correction steps need to be executed on the measured data in order to extract G_p/ω.

Figures 2.16 and 2.17 show a very good agreement between the energy profiles of the interface trap density and time constants obtained using the

58

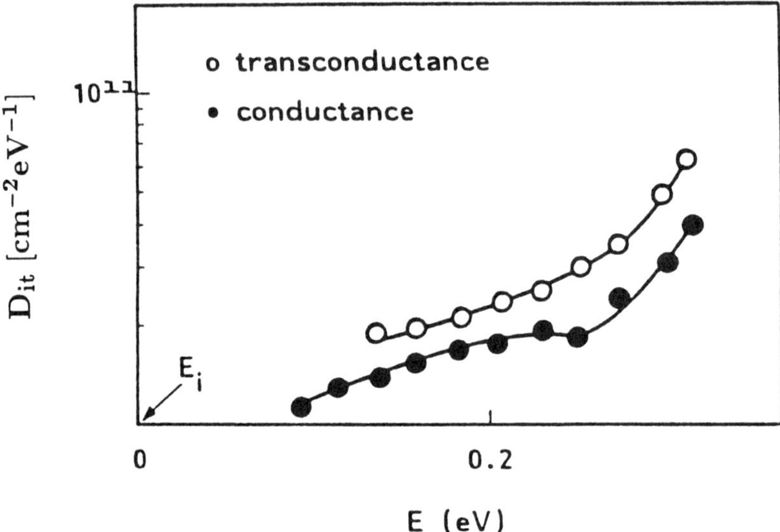

Figure 2.16: Comparison between the interface trap density profiles obtained using the conductance technique (filled circles) and the dynamic transconductance method (empty circles).

present G_p/ω method and the split-admittance technique [33]. This dynamic-transcondutance-based method has been used to characterize enhancement mode MOSFETs fabricated on partially depleted SIMOX substrates [34]. The results of this study have revealed the presence of bulk traps due to defects arising after the back gate oxide growth by ion implantation. Such traps are identified by the occurrence of a conductance peak whose position and magnitude are gate voltage independent in addition to normal peaks attributed to interface trap loss. Such double peak curves are shown in Figure 2.18.

Furthermore, the G_p/ω method has been adapted to fully depleted enhancement SOI devices [36]. It has been shown that the energy range in which the front interface is characterized can be extended by performing *direct* and *cross* dynamic transconductance measurements [37].

In case of direct measurement, the front interface is biased in weak inversion and the back gate in accumulation. The drain current flows at the front interface and G_p/ω is the same as for bulk Si MOSFETs.

Cross measurement employs the coupling of the two interfaces to explore G_p/ω at the front interface by measuring the dynamic transconductance of the back interface. In such case, the front interface is kept in depletion or accumulation whereas the back gate is biased in inversion and G_p/ω for the

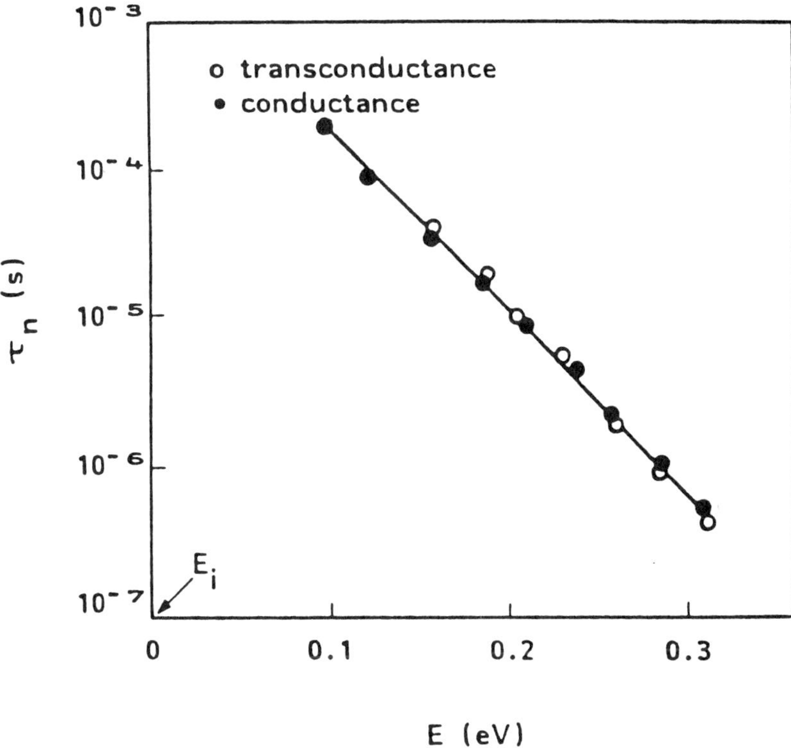

Figure 2.17: Comparison between the interface trap time constants obtained using the conductance technique (filled circles) and the dynamic transconductance method (empty circles).

front interface is given by [37],

$$\frac{G_p}{\omega} = \mu_{eff} \left(\frac{W}{L} \right) C_{ob} C_b V_d \, \Im\{g_m^{-1}(\omega)\} \tag{2.49}$$

where C_{ob} is the back gate oxide capacitance. By combining the results of the direct and cross measurements, most of the energy gap can be profiled.

The High/Low Frequency Transconductance Method

This method which has been proposed in 1991 [35], exploits the real part of the dynamic transconductance to extract the interface trap capacitance C_{it}. At high frequency (say 1 MHz), both the capacitive and conductive components of the parallel interface trap admittance go to zero, whereas at low frequency, only G_{pn} will vanish while C_{pn} will reduce to C_{it}. Consequently, C_{it} can be

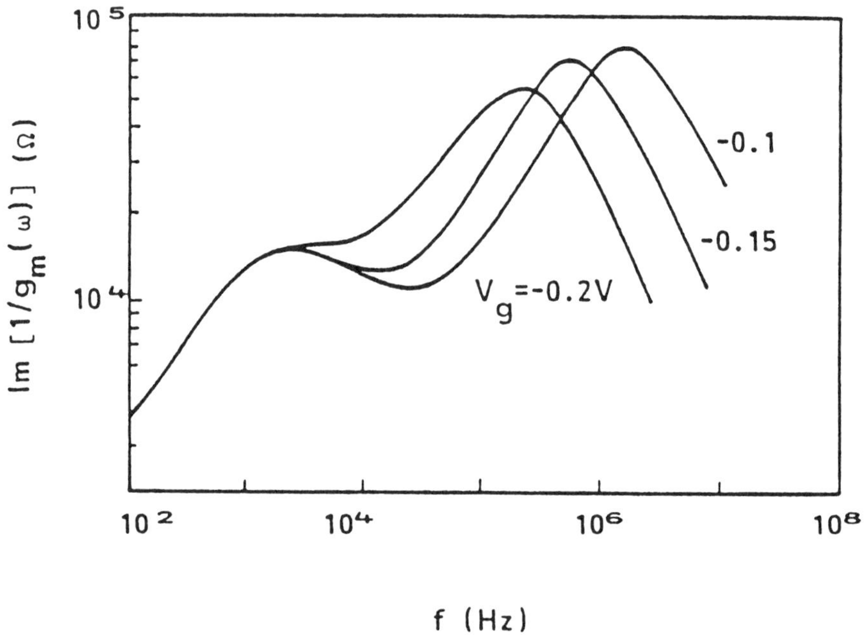

Figure 2.18: Double peak curves for $\Im(g_m^{-1})$ versus gate bias obtained on partially depleted enhancement SIMOX MOSFETs.

deduced as follows

$$C_{it} = \frac{I_d C_{ox}}{kT/q} \, \Re \left\{ \frac{1}{g_m^{LF}} - \frac{1}{g_m^{HF}} \right\} \qquad (2.50)$$

where g_m^{LF} and g_m^{HF} refer to the transconductance measured at low and high frequencies respectively.

Figure 2.19 shows the double plateau behavior of the real part of $g_m(\omega)$ versus frequency. At high frequency, the interface traps are unable to respond to the gate signal and the transconductance increases. Such behavior has been reported as early as 1965 by Becke et al [28], but it was not before 1991 when C_{it} has been elegantly expressed by Equation 2.50 [35].

In order to construct an energy profile of the interface trap density, the rate of variation of the surface potential with gate bias was expressed in terms of the low frequency transconductance as follows

$$\frac{d\psi_s}{dV_g} = \frac{kT}{qI_d} \Re\{g_m^{LF}\} \qquad (2.51)$$

The function $\psi_s(V_g)$ can be deduced by integrating the above equation where the integration constant can be evaluated by substituting with the threshold condition (*i.e.* $V_g = V_t$ at $\psi_s = 2\phi_f$) [35].

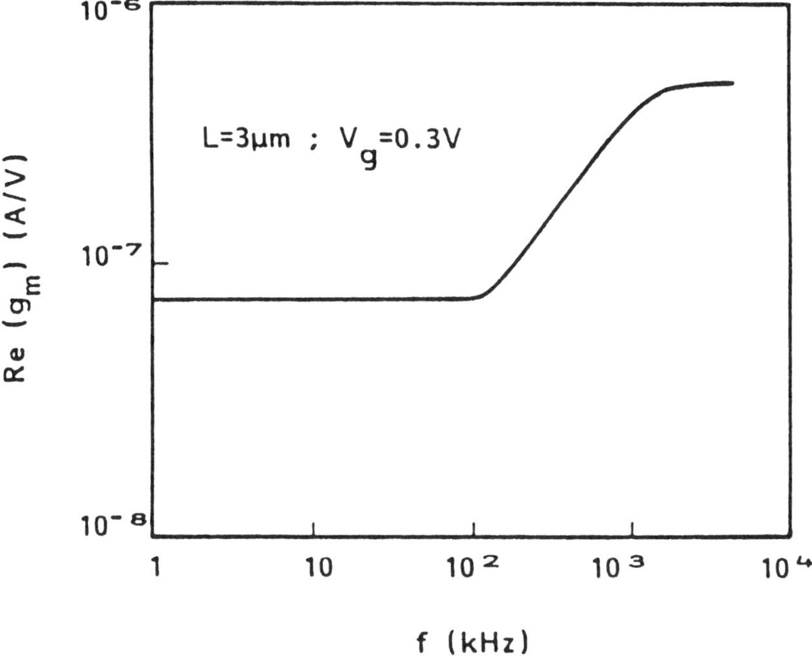

Figure 2.19: Typical variation of the real part of the dynamic transconductance versus frequency in weak inversion.

It is clear that the high/low frequency transconductance method bears resemblance to conventional capacitance methods known on MOS diodes [8], whereas the G_p/ω method is the counterpart of the classical conductance technique. The later is more sensitive than the former because the measured quantity (*i.e.* $\Im\{g_m^{-1}(\omega)\}$) is directly proportional to the interface trap loss.

2.5.2 Depletion MOSFET Dynamic Transconductance

Let us consider an n^+nn^+ depletion mode MOS transistor fabricated on an SOI film of thickness w_f, that is not totally depleted.

In the depletion regime, current conduction is carried out by majority carriers flowing in the volume of the Si film and not at the $Si - SiO_2$ interface. Therefore, the drain current for small values of drain bias can be written as

$$I_d = \mu \frac{W}{L}(Q_{df} - Q_d)V_d \qquad (2.52)$$

where $Q_{df} = qN_Dw_f$ is the maximum depletion charge per unit area, Q_d the depletion charge density, N_D the effective film doping concentration, and μ the bulk carrier mobility. As the mobility is almost depth independent in good

62

quality SOI films [39], a gate voltage independent mobility is assumed. The transconductance is therefore given by

$$g_m = \left(\mu \frac{W}{L} V_d\right) \frac{dQ_d}{d\psi_s} \frac{d\psi_s}{dV_g} \tag{2.53}$$

Following the same approach as in enhancement type MOSFETs, the dynamic transconductance can be deduced in the following form

$$g_m = K \frac{C_d C_{ox}}{C_{ox} + C_d + Y_{it}/j\omega} \tag{2.54}$$

where $K = \mu(W/L)V_d$ and C_d is the depletion capacitance per unit area. Now, based on the above model, we will present two methods that can provide together a detailed characterization of depletion mode MOS transistors.

The G_p/ω Method for Interface Trap Characterization

As pointed out earlier, the dynamic transconductance model has been developed in the depletion regime only. It is important to note that, although the same current equation is valid in inversion, the dynamic transconductance model itself will not be straightforward. This is because minority carriers, being slowly supplied by the substrate, will exhibit a certain time delay before they can reach the surface to form the inversion layer. Consequently, the inversion layer cannot be simply modeled by a pure capacitance C_i. The G_p/ω method will therefore be used in depletion mode transistors exclusively in the depletion bias region, bearing in this respect a clear analogy to the conductance technique in MOS capacitors.

As can be seen from Equation 2.54, G_p/ω is simply obtained from the imaginary part of g_m^{-1} as was the case with enhancement MOSFETs,

$$\frac{G_p}{\omega} = K C_d C_{ox} \, \Im\{g_m^{-1}(\omega)\} \tag{2.55}$$

The value of C_d at a given gate bias can be determined from the high frequency transconductance discussed in the following section.

The High Frequency Transconductance

At frequencies above 1 MHz not only the interface traps cannot follow the gate signal but also the inversion capacitance C_i vanishes. In fact, C_i disappears even at low frequencies in the order of 10–100 Hz [38]. This is similar to the case of MOS capacitors and contradictory to that of enhancement MOSFETs where minority carriers are rapidly supplied by the source and drain regions. Therefore, the high frequency transconductance g_m^{HF}, in the depletion and inversion regimes, is given by

$$g_m^{HF} = K \frac{C_{ox} C_d}{C_{ox} + C_d} \tag{2.56}$$

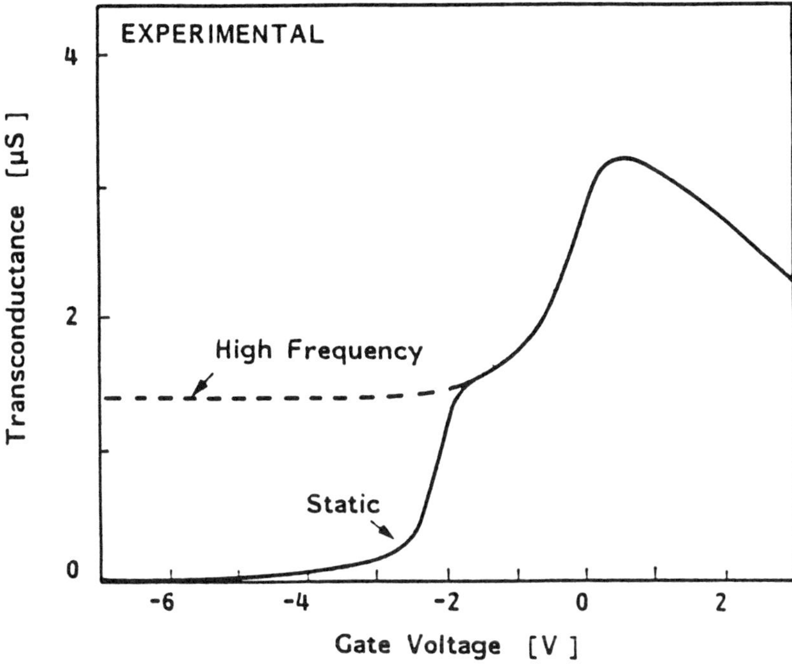

Figure 2.20: Static and high frequency transconductance characteristics for a depletion-mode SIMOX MOSFET with $L = 7\ \mu m$.

As can be seen, g_m^{HF} is nothing but the high frequency capacitance of the MOS system multiplied by an amplifying factor K ($10^2 - 10^4$ according to geometry). Figure 2.20 shows the static and high frequency transconductances measured for a SIMOX transistor of $L = 7\ \mu m$. Notice that the transconductance is an amplification of the gate-bulk capacitance of the MOS structure. In case of high frequency, it saturates at a finite minimum value in inversion just like the high frequency capacitance in an MOS diode. For very low frequencies (static curve), the transconductance drops abruptly to zero in inversion due to the existence of C_i and the transconductance has the same form as the gate-bulk capacitance C_{gb}. Notice also that in accumulation, Equation 2.54 is no longer valid as the mobility starts to decay due to scattering mechanisms.

The high frequency transconductance offers the same information for a depletion MOSFET as does the high frequency capacitance in a large area MOS capacitor due to the amplifying coefficient K. Not only g_m^{HF} is easily measurable but also the influence of parasitic capacitances is eliminated as the drain current is being measured instead of the gate capacitance.

Making use of the depletion approximation of C_d as a function of V_g [38],

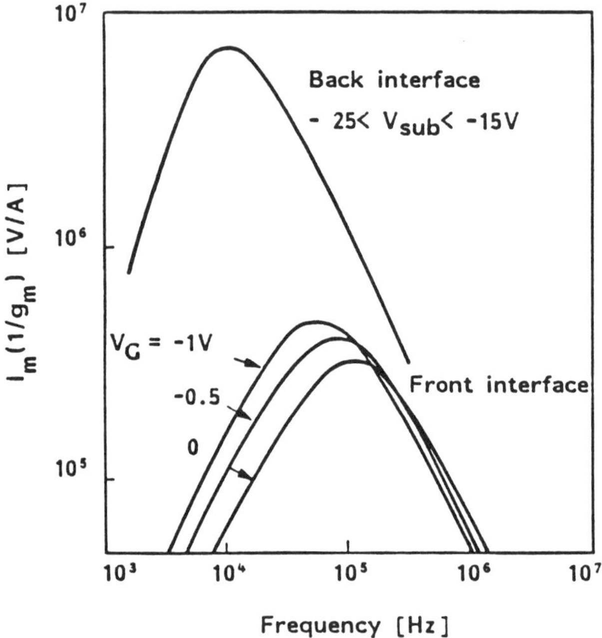

Figure 2.21: Experimental plot of $\Im(g_m^{-1})$ versus frequency for the front and back interfaces of a depletion-mode SIMOX MOSFET biased in depletion.

the high frequency g_m can be related to the MOS capacitance as follows

$$\left(\frac{g_{mmin}}{g_m^{HF}}\right)^2 = \left(\frac{C_{min}}{C_{ox}}\right)^2 + \frac{2C_{min}^2}{\epsilon_{si}qN_D}(V_g - V_{fb}) \tag{2.57}$$

where C_{min} is the series combination of C_{ox} and the minimum depletion capacitance. Therefore, by plotting the left hand side of the above equation versus gate voltage, the doping concentration is readily obtained from the slope, the mobility from the ratio $g_{mmin}/C_{min} = K$ and the flat band voltage from the intercept with the gate bias axis.

Figure 2.21 shows the results of $g_m(\omega)$ measurements on the front and back interfaces of a SIMOX transistor biased in depletion for several values of V_g [34]. As for the front interface, the obtained G_p/ω peaks vary in position and magnitude with V_g which is characteristic of interface trap loss. For the back interface, however, the observed peak is gate voltage independent and is attributed to bulk traps present in the depletion layer of the back gate.

Bibliography

[1] C. N. Berglund, *IEEE Trans. on Elect. Dev.*, ED-13, 701 (1968).

[2] R. Castagné, *C.r. Hebd. Séanc. Acad. Sci.*, Paris, 267 B, 886 (1968).

[3] M. Kuhn, *Solid St. Electron.*, 13, 873 (1970).

[4] D. R. Kerr, *Int. Conf. on Properties and Use of MIS Structures*, Grenoble, p. 303 (1969).

[5] M. Kuhn and E. Nicollian, *J. Electrochem. Soc.*, 118, 370 (1971).

[6] P. V. Gray and D. M. Brown, *Appl. Phys. Lett.*, 8, 31 (1969).

[7] E. Arnold, *IEEE Trans. on Elect. Dev.*, ED-15, 1003 (1968).

[8] E. Nicollian and A. Goetzberger, *Bell Syst. Tech. J.*, 46 1055 (1967).

[9] E. Nicollian and J. R. Brews, *MOS Physics and Technology*, Wiley, New York (1982).

[10] H. Haddara and M. Elsayed, *Solid St. Electron.*, 16, 801 (1988).

[11] J. Koomen, *Solid St. Electron.*, 16, 801 (1973).

[12] M. Elsayed and H. Haddara, *Proc. of ESSDERC'89*, Berlin, p. 940 (1989).

[13] M. Elsayed and H. Haddara, *Solid St. Electron.*, 34, 173 (1991).

[14] H. Haddara and G. Ghibaudo, *ESSDERC,86*, Cambridge, UK, p.117 (1986).

[15] R. F. Pierret, *Solid St. Electron.*, 11, 253 (1968).

[16] V. Lieneweg, *Solid St. Electron.*, 23, 577 (1980).

[17] R. Castagné, *Thesis Faculté des Sciénces d'Orsay*, Paris (1970).

[18] H. Deuling, E. Klausmann and A. Goetzberger, *Solid St. Electron.*, 15, 559 (1972).

[19] P. D. Chow and K. L. Wang, *Solid St. Electron.*, 29, 1005 (1986).

[20] S. Bassiouni, H. Haddara and H. F. Ragaie, *Solid St. Electron.*, 36, 741 (1993).

[21] Y. Tsividis, *Operation and Modeling of the MOS Transistor*, McGraw Hill, New York, 1987.

[22] J. Brugler and P. Jespers, *IEEE Trans. on Elect. Dev.*, ED-16, 297 (1969).

[23] G. Groeseneken, H. Maes, N. Beltran and R. De Keersmaeker, *IEEE Trans. on Elect. Dev.*, ED-31, 42 (1984).

[24] A. G. Sabnis and J. T. Clemens, *IEDM Tech. Dig.*, p. 18 (1979).

[25] S. C. Sun and J. D. Plummer, *IEEE Trans. on Elect. Dev.* ED-27, 1497 (1980).

[26] M. -S. Liang, J. Choi, P. -K. Ko and C. Hu, *IEEE Trans. on Elect. Dev.*, ED-33, 409 (1986).

[27] C. Sodidni, T. W. Ekstedt and J. Moll, *Solid St. electron.*, 25, 833 (1982).

[28] H. Becke, R. Hall and J. White, *Solid St. Electron.*, 8, 813 (1965).

[29] J. Koomen, *Solid St. Electron.*, 17, 321 (1974).

[30] L. Schrader, *Solid St. Electron.*, 20, 671 (1977).

[31] A. Van calster, *Solid St. Electron.*, 21, 393 (1978).

[32] H. Haddara and G. Ghibaudo, *Solid St. electron.*, 31, 1077 (1988).

[33] H. Haddara and M. Elsayed, *Proc. of ESSDERC'87*, Bologna, Italy, p. 127 (1987).

[34] H. Haddara, M. T. Elewa and S. Cristoloveanu, *IEEE Electron Dev. Lett.*, EDL-9, 35 (1988).

[35] H. -S. Chen and S. S. Li, *IEEE Electron Dev. Lett.*, EDL-12, 13 (1991).

[36] D. E. Ioannou, X. Zhang, G. Ccampisi and H. Hughes, *Proc. of the IEEE SOI/SOS Conf.*, p. 42 (1990).

[37] X. Zhang, S. Gulwadi, D. E. Ioannou, G. Campisi and H. Hughes, *Proc. of the IEEE SOI/SOS Conf.*, p. 78 (1991).

[38] S. M. Sze, *Physics and Technology of Semiconductor Devices*, Wiley, New York, (1982).

[39] S. Cristoloveanu and S. S. Li, *Electrical Characterization of Silicon On Insulator Materials and Devices*, Kluwer Academic Publishers, 1995.

3

Charge Pumping

C. R. Viswanathan
University of California at Los Angeles
California, USA

3.1   Introduction

The interface between the silicon substrate and the gate oxide in the active region of a MOS transistor plays a crucial role in determining the device parameters and affects reliability and lifetime of the device. The measurement and characterization of the interface properties are needed to understand the origin and physical properties of the interface states in an MOS system. Interface states can capture and emit charge carriers and the amount of charge in the interface states determines the device parameters. Interface state measurements are routinely carried out on test devices to monitor the manufacturing process as well as to study the lifetime and reliability of the devices. The interface measurements are also used in failure analysis investigations. The usual technique to study interface characteristics is to use a large area MOS capacitor with the gate oxide grown at the same time or in similar circumstances as when the gate oxide in the product device is grown. The Capacitance-Voltage (C-V) characteristics of the capacitor are analyzed to obtain the energy distribution of fast interface state density, fixed oxide charge and slow interface state density. Fast interface states are those which are located just at the interface and therefore can readily exchange charge with the substrate by capturing and emitting charge carriers. The amount of charge residing in the fast interface states is a function of the surface potential or band-bending and therefore a function of the gate voltage. The slow interface states are those that are located in the oxide within a short distance from the interface so that the charge carriers can tunnel through the oxide between the slow states and the substrate. The slow states can charge up and discharge with a time constant equal to the tunneling time constant. Since the tunneling time constant increases exponentially with the tunneling distance, only states which are very close to the interface will participate in charge exchange in a reasonable amount of time. The difficulty with the C-V technique using a MOS capacitor is not only the requirement of a large area device but also the problem of extrapolating the results of the measurements to the actual device in the product line. Another technique that is used to investigate the interface characteristics is the Deep Level Transient Spectroscopy (DLTS). DLTS measurements require an elaborate experimental set-up using synchronized pulse sources, integrators and time averagers. DLTS is not suitable for routine testing. Furthermore DLTS also requires a large area test device and cannot be carried out directly on a short channel MOS transistor.

The charge pumping technique [1] has evolved into a sensitive and reliable method to study the interface characteristics in a MOS transistor. The major advantage of the charge pumping (CP) technique is that the measurements are done on the actual transistor itself without needing a separate test device. The need to extrapolate the results obtained on a test device to characterize the interface in a transistor is thus eliminated. Very few techniques other than charge pumping are available for direct measurement of the interface characteristics in an MOS transistor. The technique yields precise results. The technique can be used to obtain quick results in a simple experiment or to obtain detailed results on the interface states by sophisticated modification of the basic CP experiment. The energy distribution as well as the lateral distribution of the interface states in the channel can be determined. The sensitivity of the CP technique permits the measurement on small geometry transistors present in modern VLSI technology. In short channel devices, the channel is non-uniform and the CP technique enables the probing of the interface as a function of position in the non-uniform channel. The CP technique is widely used to study defects induced in short channel devices by hot carrier stress. Degradation of the short channel device under hot carrier stress places a severe limitation on the lifetime and reliability of short channel devices. In a device that has been subjected to hot carrier stress, the interface damage is localized to a region in the vicinity of the drain. The profiling methods developed using the CP technique permit the investigation of the damaged interface regions. The CP technique is also useful in studying the interface damage in devices that have been subjected to Fowler-Nordheim (F-N) stress or exposed to radiation. The CP technique has also been used to study Silicon-on-Insulator (SOI) MOS devices.

The only drawback of the CP technique is the difficulty in the modeling of the CP process when non-steady state conditions exist during the rise and fall times of the gate pulse. But using simplifying assumptions, the effect of finite rise and fall times has been modeled and the technique has been successfully employed to obtain capture cross section parameters for electrons and holes. By combining two dimensional simulation programs with CP results, the interface state characteristics in short channel devices can be investigated in great detail.

In this chapter, the principles and applications of the charge pumping technique are discussed. Various modifications of the technique that have been used by different groups of investigators are described.

3.2 Early Experiments and Basic Principle of CP Measurement

The first experiment on charge pumping was done in 1969 by Brugler and Jespers [1]. Elliot [2] demonstrated that the charge pumping technique can be successfully employed to study interface characteristics. The principle of charge pumping can be understood by reference to Figure 3.1. An MOS transistor has a pulse voltage source connected between the gate and the ground. The substrate is grounded through a dc ammeter. The source and drain are connected together and a reverse voltage is applied with respect to ground. The pulse switches the surface between accumulation and inversion conditions. Unless otherwise stated, we will be using an nMOS transistor as an example in our discussion of the charge pumping phenomenon. The charge pumping action proceeds in the following manner. When the pulse (of sufficient amplitude to invert the surface) is on, electrons flow to the surface from the source and the drain. The interface states capture the inversion electrons and become negatively charged. When the pulse is off, the inversion layer electrons immediately flow back to the source and the drain, while the captured electrons are still retained by the interface states. However, the surface is accumulated now and the holes in the accumulation region are captured by the negatively charged interface states resulting in a recombination of the electrons captured during the ON period and the holes captured during the OFF period. The holes needed in this recombination flow through the substrate. Thus charge is pumped from the source and drain to the substrate via the interface states in each cycle of the pulse. Hence the process is called "Charge Pumping".

The amount of charge pumped per cycle is equal to the electron charge times the number of interface states capturing electrons and holes alternately. In Figure 3.2(a), the location of the Fermi level in the bandgap at the surface is indicated during the pulse OFF condition by $E_F{}^{OFF}$ and during the pulse ON condition by $E_F{}^{ON}$. As the gate is switched back and forth between accumulation and inversion, the Fermi level switches between these values. Denoting the corresponding surface potential (or band bending) as Ψ_S^{OFF} and Ψ_S^{ON}, the change in the surface potential between ON and OFF conditions is equal to

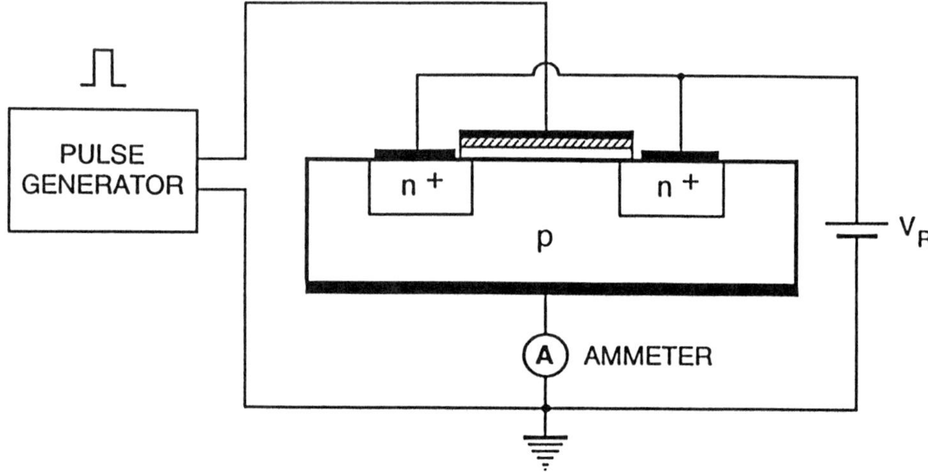

Figure 3.1: Basic experimental arrangement for Charge Pumping measurement

$\Delta\Psi_S$ given by

$$\Delta\Psi_S \;=\; \Psi_S^{ON} - \Psi_S^{OFF} \;=\; \frac{E_F^{ON} - E_F^{OFF}}{q} \tag{3.1}$$

Figure 3.2(b) shows an arbitrary distribution of the interface states in the bandgap at the surface. The interfaces states lying between E_F^{ON} and E_F^{OFF}, shown as the shaded region in this figure, alone will participate in the charge pumping process. The amount of charge pumped per cycle, Q_{CP}, is equal to

$$Q_{CP} \;=\; qA \int_{E_F^{OFF}}^{E_F^{ON}} D_{it}(E)dE \;=\; qA\overline{D_{it}}\Delta\Psi_S \tag{3.2}$$

where $D_{it}(E)$ is the energy distribution of interface states, $\overline{D_{it}}$ is the average interface state density between Ψ_S^{ON} and Ψ_S^{OFF} and q is the electron charge. Due to repeated charge pumping during each pulse cycle, unidirectional pulses of charge flow through the substrate and a dc substrate current flows equal to

$$I_{SUB} \;=\; Q_{CP}f \;=\; qAf \int_{E_F^{OFF}}^{E_F^{ON}} D_{it}(E)dE \;=\; qAf\overline{D_{it}}\Delta\Psi_S \tag{3.3}$$

where f is the repetition frequency of the gate pulse. This current is usually denoted I_{CP} rather than I_{SUB}. Typically it is assumed that in strong inversion, the band bending is pinned to a value equal to twice the Fermi potential Φ_F,

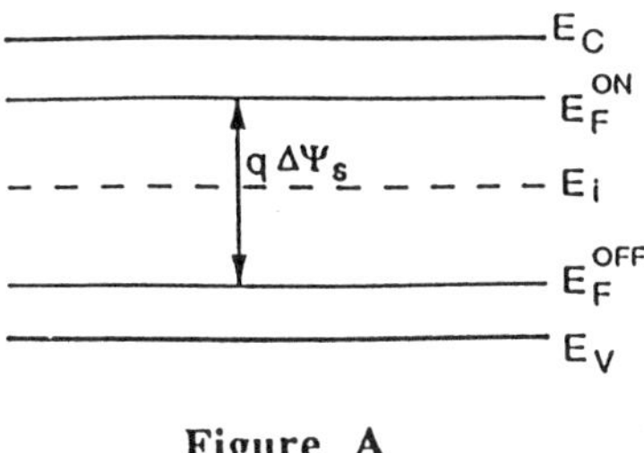

Figure A

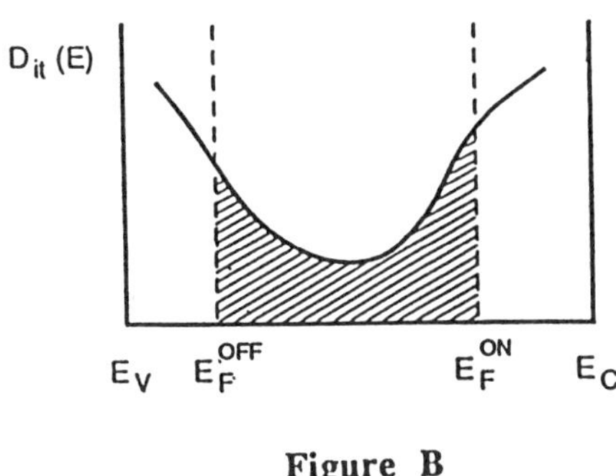

Figure B

Figure 3.2: Illustration of the band of interface states participating in CP action; (a) The location of the Quasi-Fermi level during the ON and OFF period of the pulse. (b) An arbitrary distribution of the interface states in the bandgap. The shaded region represents the interface states contributing to the CP current.

and that in accumulation, the band bending is nearly zero. This makes $\Delta\Psi_S$ in the above equation equal to $2\Phi_F$. By measuring I_{CP}, one can, in principle, obtain information on the interface state density.

In the description above, several simplifying assumptions have been made. One is that, when the pulse is turned off, all the mobile (inversion) carriers flow back to the source and the drain region. This does not happen always. When the length of the channel is large, the carriers in the middle of the channel, which are far away from the source and drain junctions, will not have had time to flow back to the source and drain by the time the pulse is turned off. These carriers that are left behind will recombine with majority carriers (holes) during the OFF period and therefore an additional component of charge pumping that does not involve the interface states arises. The dc substrate current resulting from the recombination of inversion (mobile carriers) with majority carriers in the substrate is an unwanted current. Since the number of carriers left behind in the channel depends on the length and the width of the channel, this dc

72

substrate current is called the geometric component. In the charge pumping experiment both the interface charge pumping and the geometric components are measured. If α is the fraction of the inversion carriers that do not flow back to the source and the drain and then recombine in the substrate, the charge pumping current including the interface and geometric components can be expressed as

$$I_{CP} = fA \left[q\overline{D_{it}}\Delta\Psi_S + \alpha C_{ox}(V_{GH} - V_T) \right] \tag{3.4}$$

where C_{ox} is the gate capacitance per unit area, V_T the threshold voltage and V_{GH} the gate voltage at the top or high value of the pulse. The geometric component interferes with the use of the charge pumping technique for interface state density measurements. Several techniques are employed to reduce the geometric component. One is to apply a larger reverse voltage between the substrate and the source/drain regions. This technique has the disadvantage of altering the effective channel length which has to be determined by auxiliary measurements. It also alters the threshold voltage in the channel. Some authors [3, 4] have claimed that increasing the W/L ratio reduces the geometric component. Some others [1, 5] have claimed that only reducing the length of the channel decreases the geometric component. Using transistors with a larger W/L ratio, while keeping L constant, does not alter the fraction of the measured charge pumping current that can be attributed to the geometric component. Increasing W/L ratio by decreasing L reduces the time needed for the inversion carriers in the middle of the channel to reach the source/drain region. The technique used in the earlier days to minimize the geometric component was to increase the fall time of the gate pulse. Under this condition, the inversion carriers are able to empty completely into the source/drain region by the time the gate is turned off.

A typical pulse shape used in charge pumping experiment is shown in Figure 3.3. V_{GL} is the voltage at the bottom of the pulse and V_{GH} is that corresponding to the top of the pulse. In order for the charge pumping experiment to work as stated earlier, V_{GL} has to be less than the flatband voltage, V_{FB}, in order to correspond to accumulation condition and V_{GH} has to be larger than the threshold voltage V_T in order to correspond to inversion condition.

Till now, the discussion has assumed that the surface potential is uniform everywhere in the channel for a given gate voltage. In reality, the band bending is more near the source and the drain and less in the middle of the channel under depletion and inversion conditions. This means that as the gate voltage is increased gradually, inversion occurs first at the two ends of the channel (under conditions of zero drain to source voltage) and spreads towards the middle of the channel at higher gate voltages. While one can assume an equipotential surface over a large portion of the mid-region of the channel in long channel devices, the surface is not equipotential even in the middle of the channel in short channel devices. The surface potential changes with gate voltage in inversion and in accumulation albeit by a small amount. The expression for the charge pumping current that avoids these simplifications can be developed as follows. Denoting

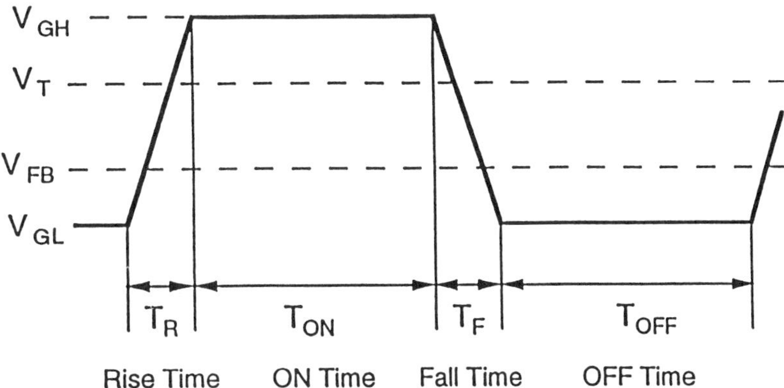

Figure 3.3: The shape of the gate pulse used in CP measurements. The finite rise and fall times, T_R and T_F respectively, of the pulse make it a trapezoidal pulse. T_{ON} and T_{OFF} are the ON and OFF periods of the pulse respectively.

the surface potential at the low level of the pulse voltage, V_{GL}, as Ψ_S^{OFF}, and that at high level of the pulse voltage V_{GH} , as Ψ_S^{ON}, and recognizing that Ψ_S^{OFF} and Ψ_S^{ON} both are functions of y, the position in the channel measured from the source, the charge pumping current can be expressed as

$$I_{CP} \;=\; qWf \int_0^L \left[\int_{\Psi_S^{OFF}(y)}^{\Psi_S^{ON}(y)} D_{it}(\Psi_s)d\Psi_s \right] dy \tag{3.5}$$

where W and L are the width and the length of the channel. Implicit in this expression is the assumption that $\Psi_S^{ON} > 2\Phi_F$ everywhere in the channel *i.e.* the entire channel is inverted. At a gate voltage slightly less than V_T, the channel will be inverted only in a limited region of length ΔL near the source and similarly near the drain as shown in Figure 3.4. In such a case, the expression for the charge pumping current has to be modified as

$$I_{CP} \;=\; I_{CPS} + I_{CPD} \tag{3.6}$$

where

$$I_{CPS} \;=\; qWf \int_0^{\Delta L} \left[\int_{\Psi_S^{OFF}(y)}^{\Psi_S^{ON}(y)} D_{it}(\Psi_s)d\Psi_s \right] dy \tag{3.7}$$

and

$$I_{CPD} \;=\; qWf \int_{L-\Delta L}^L \left[\int_{\Psi_S^{OFF}(y)}^{\Psi_S^{ON}(y)} D_{it}(\Psi_s)d\Psi_s \right] dy \tag{3.8}$$

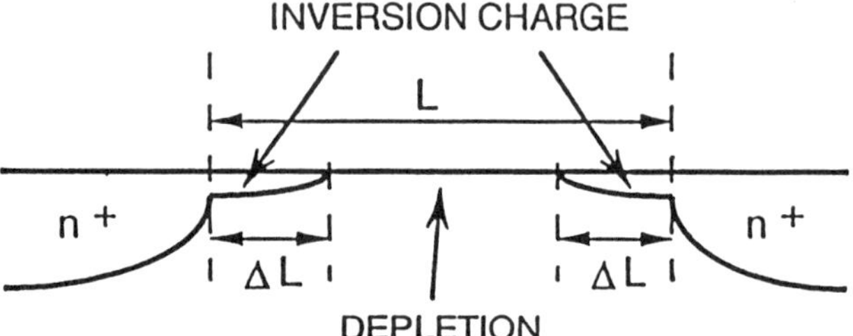

Figure 3.4: The formation of the inversion region near the source and the drain while the middle of the channel is still depleted at a gate voltage slightly less than V_T. ΔL is the length of the inverted region near the source and the drain.

I_{CPS} is the CP current flowing through the source and I_{CPD} is the current flowing through the drain when the channel is only partially inverted. As V_{GH} is increased, the length of the inverted regions at the two ends of the channel increases and the charge pumping current increases. Both these components saturate when V_{GH} exceeds the threshold voltage V_T and the entire channel is inverted as shown in Figure 3.5.

3.3 Interface State Generation-Recombination Kinetics

The derivation of Equations 3.5, 3.7 and 3.8 is based on the assumption that the capture of electrons in inversion and holes in accumulation are the only processes that the interface states are involved in. But, the interface states can change their charge state by emission processes also. It is necessary to understand the kinetics of capture and emission processes in an interface state so that the results of charge pumping can be properly analyzed. The process of capture and emission of carriers by interface states can be described using the Shockley-Read-Hall (SRH) model. According to the SRH model, an interface state can be in either a neutral state or in a singly charged state. An interface state is called a donor type state if it exists either in a neutral state or in a positively charged state. Similarly it is called an acceptor type if it exists either in a neutral state or in a negatively charged state. Interface states cannot be distinguished as acceptor or donor type states in an electrical measurement. The interface state exists in a less negative state or in a more negative state (or alternately in a more positive state or in a less positive state) independent of whether it is a donor or acceptor type of state. An interface state in a neutral state can get charged either by capturing a charge carrier of one type or by emitting a carrier of the opposite charge type. This charged interface

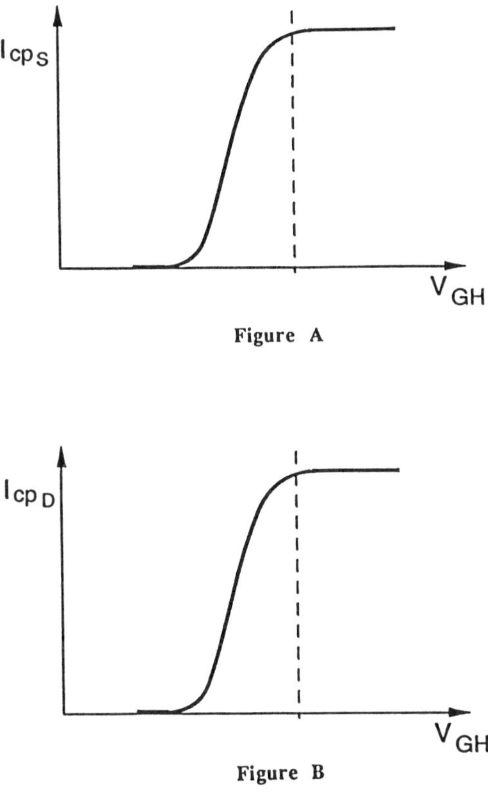

Figure 3.5: The CP current flowing through the source and drain leads. (a) I_{CPS} is the current flowing through the source lead and (b) I_{CPD} is the current flowing through the drain lead.

state can return to the neutral state either by emitting the charge carrier of the same polarity to the one it captured or by capturing a charge carrier of opposite polarity. For example, a neutral donor type state can either emit an electron or capture a hole to become positively charged while a negatively charged acceptor type state will become neutral either by emitting an electron or by capturing a hole. Similarly, the charged donor type interface state can emit a hole or capture an electron to become neutral while a neutral acceptor type atom will emit a hole or capture an electron to become negatively charged. The transition from one charge-state to another is illustrated in Figure 3.6 for an acceptor type interface state. The same diagram is valid for a donor type state if the label "neutral state" is changed to "positively charged state" and the label "negatively charged state" is changed to "neutral state".

The probability per unit time that an interface state will emit or capture a

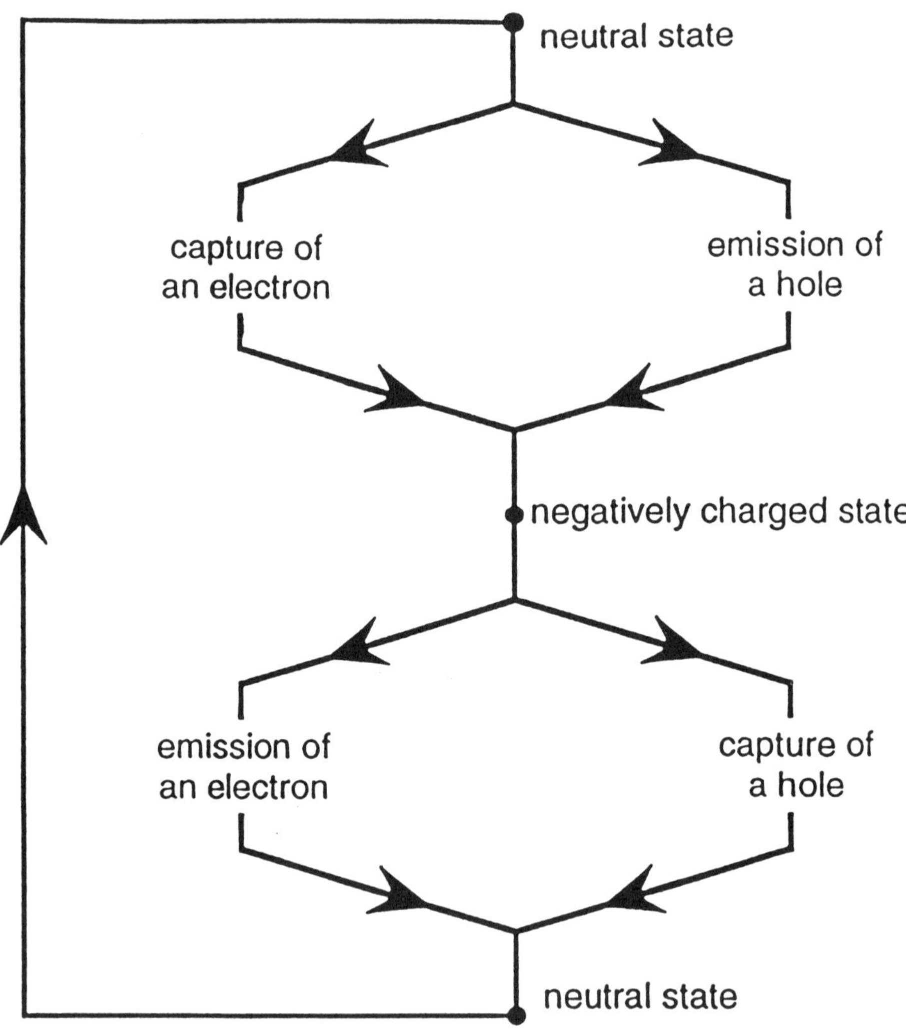

Figure 3.6: Shockley-Read-Hall (SRH) model for the capture and emission of carriers by interface states. The interface state is assumed to be an acceptor type of state in this diagram. The same diagram can be used to describe the processes in a donor type state by renaming the neutral state as positively charged state and the negatively charged state as neutral state.

charge carrier can be shown to be:

a) for electron capture: $c_n = \sigma_n v_{th} n_s$

b) for electron emission: $\epsilon_n = \sigma_n v_{th} n_1$

c) for hole capture: $c_p = \sigma_p v_{th} p_s$

d) for hole emission: $\epsilon_p = \sigma_p v_{th} p_1$

where σ_n and σ_p are electron and hole capture cross sections of the interface states; v_{th} is the thermal velocity of the charge carriers; n_s and p_s are electron and hole densities at the surface.

$$n_1 = n_i\, e^{\frac{E_t - E_i}{KT}} \tag{3.9}$$

and

$$p_1 = n_i\, e^{\frac{E_i - E_t}{KT}} \tag{3.10}$$

where E_t is the energy of the interface state in the band gap.

Independent of which type the interface state is, successive capture of an electron and a hole by an interface state initially in the neutral state returns it to the neutral state. In this process, the interface state has caused a recombination of an electron and a hole. Similarly an interface state initially in a neutral state can successively emit an electron and a hole and return to the neutral state. In this case the interface state has caused a generation of an electron and hole. If for a group of interface states, $\sigma_n \approx 0$ while $\sigma_p \neq 0$, then these states can capture or emit only holes. These states will act as hole traps. On the other hand, if $\sigma_p \approx 0$ while $\sigma_n \neq 0$, then the states can capture or emit only electrons. These states will act as electron traps. When $\sigma_n \approx \sigma_p \neq 0$, then the states will be able to interact with both types of carriers and hence give rise to both generation and recombination of electron-hole pairs. In such a case, the states act as generation$-$recombination (g-r) centers.

If N_{it} is the areal density of states (i.e. per unit area in a small interval dE), N_{it}^0 is the density of states in the neutral state and N_{it}^- is the density of states in the negatively charged states[1], it can be shown that under steady state conditions,

$$N_{it}^- = N_{it} \frac{\sigma_n v_{th} n_s + \sigma_p v_{th} p_1}{\sigma_n v_{th}(n_s + n_1) + \sigma_p v_{th}(p_s + p_1)} \tag{3.11}$$

and

$$N_{it}^0 = N_{it} \frac{\sigma_n v_{th} n_1 + \sigma_p v_{th} p_s}{\sigma_n v_{th}(n_s + n_1) + \sigma_p v_{th}(p_s + p_1)} \tag{3.12}$$

If the surface is accumulated, then $p_s >> n_s$ and assuming that the energy E_t of the interface state is above the Fermi energy, $p_s > p_1$ and n_1,

---

[1] We are assuming acceptor type of states.

$$N_{it}^{-} \approx 0 \tag{3.13}$$

$$N_{it}^{0} \approx N_{it} \tag{3.14}$$

This means that the interface states are in the neutral state. If the surface is in strong inversion then, $n_s >> p_s$ and with corresponding assumptions of E_t being below E_F, we get

$$N_{it}^{-} \approx N_t \tag{3.15}$$

and

$$N_{it}^{0} \approx 0 \tag{3.16}$$

The interface states are negatively charged. The interface states are not located at a single discrete energy level but in a continuum of states distributed in energy over the band gap as shown in Figure 3.2(b). Taking $D_{it}(E)$ as the interface state density distribution, the states below E_F^{OFF} will be negatively charged in accumulation (*i.e.* during the OFF period) and states below E_F^{ON} will be negatively charged in inversion (*i.e.* during the ON period) where E_F^{OFF} and E_F^{ON} are the Fermi energy level in accumulation and inversion respectively.

We will now make an assumption that the probability rate for an interface state to make a transition between the neutral and the charged condition is valid in both steady-state and non-steady-state conditions. Referring to Figure 3.6, the probability per unit time that a neutral state will become negatively charged is

$$c_n + \epsilon_p = \sigma_n v_{th} n_s + \sigma_p v_{th} p_1. \tag{3.17}$$

Whether the capture process or the emission process dominates will depend generally on the relative magnitudes of n_s and p_1. p_1 depends on the location of the energy level of the interface state in the band gap and n_s, the surface density of electrons, depends on the location of the Fermi energy in the band gap at the surface and therefore on Ψ_s, the surface potential.

Similarly, the probability per unit time that an interface state which is in the negatively charged state will become neutral is given by

$$c_p + \epsilon_n = \sigma_n v_{th} p_s + \sigma_n v_{th} n_1. \tag{3.18}$$

Using these expressions for probability per unit time, one can show that when surface conditions are suddenly changed to force the interface state to change the charge-state, the transition from neutral to negatively charged state occurs exponentially in time with a time constant equal to

$$\frac{1}{\sigma_n v_{th} n_s + \sigma_p v_{th} p_1}. \tag{3.19}$$

Similarly the transition from the negatively charged state to the neutral state will occur exponentially in time with a time constant equal to

$$\frac{1}{\sigma_n v_{th} n_1 + \sigma_p v_{th} p_s}. \tag{3.20}$$

The above equations take into account the presence of both capture and emission processes in the transition between neutral and charged conditions.

We are assuming that T_{ON} and T_{OFF} are sufficiently long that steady state conditions are reached and that the interface states are in equilibrium with the free carriers in the substrate during both the ON and OFF periods. In a charge pumping experiment, if the gate pulse switches the device between strong accumulation and strong inversion in a manner in which there is no time for any appreciable emission to take place, then the interface states charge and discharge only by a capture process and the amount of charge pumped per pulse cycle will be given by Equation 3.2.

Let us now assume that in a charge pumping experiment the rise time T_R of the gate pulse shown in Figure 3.3 is not very short and that some of the interface states start emitting holes during this period. Until the rising edge of the pulse reaches V_T, there are no electrons at the surface (*i.e.* $n_s \approx 0$). The emission time constant as given by Equation 3.19 is a function of p_1 only.

Using Equation 3.10 for p_1, the hole emission time constant for the interface states at energy E_t in the bandgap can be expressed as

$$\tau (E_t) \ = \ (\sigma_p \, v_{th} \, n_i)^{-1} \, e^{ - (\frac{E_i - E_t}{kT}) } \tag{3.21}$$

This shows that the interface states lying closer to the valence band (far from E_i) have a shorter emission time constant than states closer to E_i.

A simple analysis can be made of the non-steady state conditions during the rise time of the pulse. The rise time can be divided into four intervals t_1, t_2, t_3 and t_4 as shown in Figure 3.7. V_{Gi} is the gate voltage at which the surface becomes intrinsic. During the interval t_1, we will assume that the interface states are in equilibrium. We are assuming that there is no change in Ψ_s during t_1, (*i.e.* there is negligible band bending when the gate voltage is less than V_{FB}), During this period, the interface states distributed up to an energy level E_F^{OFF} are negatively charged assuming the interface states to be acceptor type ones. During the interval t_2, the hole density decreases from its value that it had during t_1 and the interface states are in non-equilibrium condition. We will assume that in the interval $t_2 + t_3$, the surface carriers, holes and electrons, are in equilibrium with the rising pulse voltage while the interface states are not in equilbrium. The interface states start emitting holes (they cannot capture electrons because the electron density is low). In a time interval t_2, the interface states up to some energy level would have emitted holes. When the voltage during the rise time becomes V_{Gi}, the electron capture starts to become significant. During the period t_3, the interface states become negatively charged by both electron capture and hole emission processes. States which have energy lower than the intrinsic Fermi energy by an amount Δ will continue to be emitting holes until the Quasi-Fermi level for surface electrons gets above $E_i + \Delta$. After the Quasi-Fermi level sweeps above $E_i + \Delta$, the electron capture will dominate. To summarize, interface states lying at an energy $E_i - \Delta$, will emit holes until the Quasi-Fermi level for surface electrons

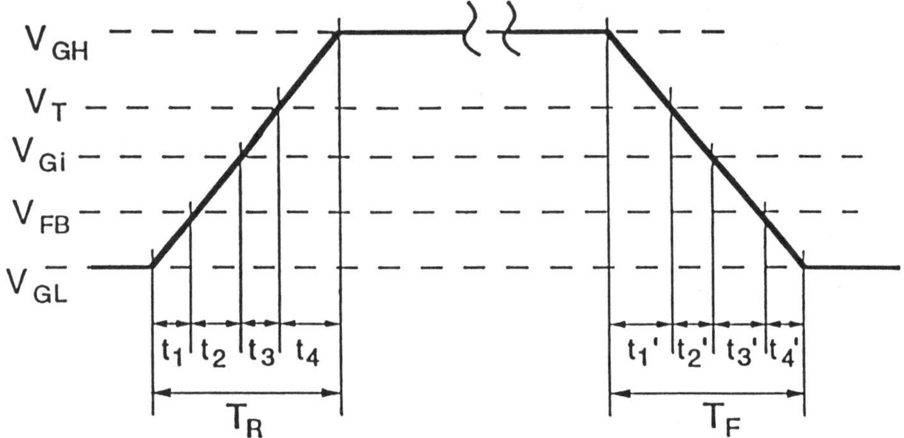

Figure 3.7: Division of the rise and fall time of the pulse into equilibrium and non-equilibrium periods. The interface states are assumed to be in equilibrium during t_1, t_4, t_1' and t_4'. During t_2, t_3, t_2' and t_3' the interface states are not in equilibrium.

crosses an energy level $E_i + \Delta$. When the Quasi-Fermi level exceeds $E_i + \Delta$, the interface state will capture an electron if it has not already emitted a hole.

When the rising pulse voltage becomes equal to V_T, the surface gets inverted. Since n_s is large, the capture process is fast and all the remaining interface states get negatively charged by capturing electrons up to an energy level E_F^{ON}. The interface states that have emitted holes during the period $t_2 + t_3$ and become negatively charged will not capture electrons and therefore do not contribute to I_{CP}. While it is straightforward to assume that during the interval t_2, emission of holes takes place, it is not so during the interval t_3, in which period both electron capture and hole emission take place. It is reasonable to assume that during the early stages of t_3, hole emission is still significant while in the later stages of t_3, electron capture becomes significant. The analysis becomes very difficult. Groeseneken $et\ al$ [3] made a simplifying assumption that during the time interval $t_2 + t_3$, only hole emission takes place and during t_4 capture of electrons occurs. Since the inversion electron density n_s is so large, it is reasonable further to assume that the capture process is completed by the end of t_4. The time interval $t_2 + t_3$ is denoted $t_{em,h}$. Among all the interface states, only those which have an emission time constant less than $t_{em,h}$ would have emitted holes by the time inversion is reached. Denoting the energy level of states having a time constant equal to $t_{em,h}$ as $E_{em,h}$, only states with energy above $E_{em,h}$ would contribute to I_{CP} and thus $E_{em,h}$ is the lower limit of the energy of the band of interface states contributing to charge pumping.

Similarly the fall time of the pulse can be divided into four intervals $t_1^{'}, t_2^{'}, t_3^{'}$ and $t_4^{'}$. As before, during the time intervals $t_1^{'}$ and $t_4^{'}$, the interface states are in equilbrium. It is only during the intervals $t_2^{'}$ and $t_3^{'}$ that the interface states are not in equilibrium. During $t_2^{'}$, as the electron density n_s decreases with the decreasing pulse voltage, the quasi Fermi level for electrons becomes increasingly lower than the value E_F^{ON}. States with energy level between E_F^{ON} and the instantaneous quasi Fermi level emit electrons. If these electrons are emitted before the inversion layer disappears, the emitted electrons reach the source and drain regions and no charge pumping action takes place. On the other hand, if the electrons are emitted after the inversion channel disappears, the electrons emitted into the substrate recombine with the majority carriers and contribute to the charge pumping current. In this sense the situation is different between the rise time and the fall time. During the fall time, only electrons emitted before the inversion channel disappears do not contribute to the charge pumping current while electrons emitted later contribute to the charge pumping current. During the rise time, all the holes that are emitted do not contribute to charge pumping current. During the time interval $t_2^{'}$, the electron density is still greater than n_i and it is reasonable to assume that electrons emitted during this period essentially reach the source and drain. On the other hand, some of the electrons emitted during $t_3^{'}$, particularly those emitted in the later part of $t_3^{'}$ would have been emitted after the inversion layer has disappeared.

Assuming that states having an energy above a certain level $E_{em,e}$ would be emitting electrons before the inversion channel disappears, $E_{em,e}$ becomes the upper limit of the energy band of the interface states contributing to the charge pumping current. As stated before, it is very difficult to do an exact analysis. To simplify the analysis, it can be assumed [3] that during $t_2^{'} + t_3^{'}$, electrons are emitted up to some energy level, $E_{em,e}$, and that these electrons reach the source and the drain and therefore do not contribute to CP current. $E_{em,e}$, defines the upper level of the band of interface states giving rise to CP current. If the rise and fall times are very short in comparison with the emission time, practically no emission takes place during either the rise or fall times and the two limits of the energy band of the interface states are E_F^{OFF} and E_F^{ON}. On the other hand if the rise and fall times are sufficiently long so as to allow for appreciable emission of carriers by interface states, then the two limits of the energy band of charge pumping states are $E_{em,h}$ and $E_{em,e}$.

To summarize, the band of states that are switched between two charge states (neutral and negatively charged for acceptor type of interface states) that contribute to the CP process, depends on the rise and fall times of the pulse as well as on its base and top levels. We have assumed an acceptor type of interface states in this discussion. The discussion is exactly the same for donor type of states also, since the polarity of the change in charge in the interface states between accumulation and inversion is the same for either type. We have also assumed an n-channel MOS transistor in our discussion. The same discussion can be applied to a p-channel transistor by taking into account

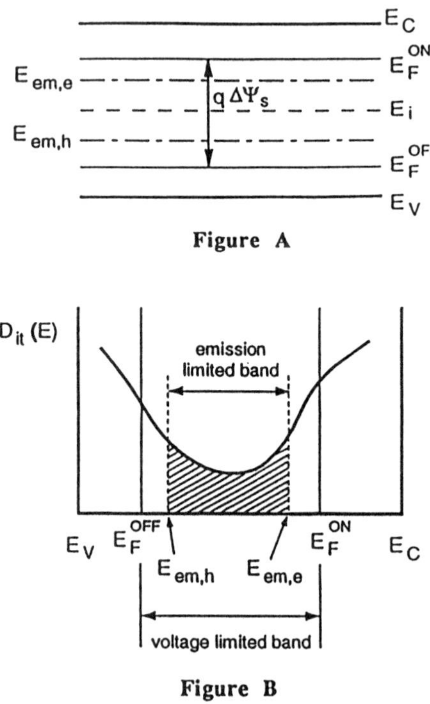

Figure 3.8: The effect of carrier emission during the rise time and fall time. The band of interface states that contribute to CP current is determined by $E_{em,h}$ and $E_{em,e}$.

the reversal of polarity of charge, voltage and current in a p-channel device from that in an n-channel device.

3.4 Experimental Techniques

The charge pumping measurement is usually carried out using rectangular pulses in either a variable amplitude mode or a variable base mode. The rectangular pulses are also called trapezoidal pulses due to the finite rise and fall times.

3.4.1 Variable Amplitude Mode

In the variable amplitude mode, the base level of the pulse, V_{GL}, is fixed while the charge pumping current is measured as a function of the amplitude (*i.e.*

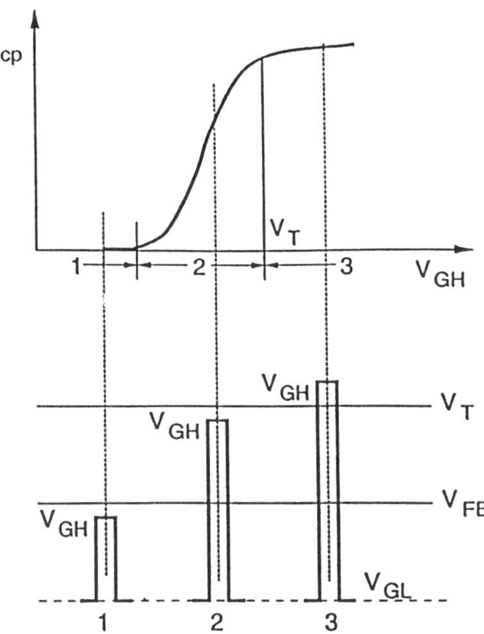

Figure 3.9: The CP current characteristics in the variable amplitude mode. V_{GH} is varied keeping V_{GL} fixed corresponding to accumulation. V_T is the threshold voltage when the entire channel is inverted.

$V_{GH} - V_{GL}$) by varying V_{GH}. A typical charge pumping current characteristic is shown in Figure 3.9. Here V_{GL} is kept below the flatband voltage V_{FB} such that the device is in accumulation during the OFF period of the pulse. The current nearly saturates when V_{GH} exceeds V_T.

Three distinct regions of the characteristics can be observed by reference to Figure 3.9. In region 1, V_{GH} is less than V_{FB}. Hence the surface is in accumulation during both ON and OFF periods. No charge pumping current is observed in this region. In region 3, V_{GH} is larger than V_T. Hence as discussed in the earlier section, the charge pumping current is due to the charging and discharging of interface states between inversion and accumulation. These states are distributed typically over an energy range of width $q\Delta\Psi_S \approx q(2\Phi_F)$ in which case the CP process can be called voltage limited CP or of width $(E_{em,e} - E_{em,h})$ in which case the process can be called emission limited CP. The fact that the current nearly saturates in this region validates our assumption that the Fermi potential is pinned in both inversion and accumulation. The flat portion has a very small slope. In the voltage limited CP case, the surface potential increases very slowly with V_{GH} in inversion and hence the interface states that participate in charge pumping increase by a small amount for an incremental change in

V_{GH}. It must be noted that the slight increase in current with V_{GH} in this region can be used for determining the interface state density in the region of the band gap close to the conduction band from a knowledge of Ψ_S variation with V_G in strong inversion. In the emission limited CP case, the slight increase in I_{CP} with V_{GH} can also be due to the fact that the time for emission becomes smaller as V_{GH} is increased keeping the rise and fall time constant. Ignoring the small increase in the current with V_G in this region, this current is taken as the maximum charge pumping current I_{CPmax}.

In region 2, V_{GH} is less than V_T. The current is less than I_{CPmax} for several reasons. One is that although V_{GH} is less than V_T, there is inversion close to the source and to the drain (Figure 3.4) as discussed earlier. Charge pumping current arises due to the interface states in this region. By studying the characteristics during the rising portion of the current characteristics, the non-uniformity of the threshold voltage can be investigated. If the channel is uniform as in a long channel device, where the threshold voltage V_T is the same over most of the channel, there is still a finite slope to the rising portion of the charge pumping characteristics. The surface is in weak inversion in the region of the rising edge of the I_{CP} characteristics. The current in this region is determined by capture of electrons in weak inversion during the ON period. The measured current is less than what is obtained in strong inversion for two reasons. One is that the portion of the band gap over which the Fermi energy changes between the ON and OFF periods is smaller and hence the number of interface states contributing to the charge pumping action is less. Second is that due to the competing hole emission process, some of the interface states, especially those with a short hole emission time constant, become charged by hole emission process rather than by electron capture process. Hence these interface states do not contribute to the charge pumping current.

3.4.2 Variable Base Mode

In the variable base mode, initially proposed by Elliot [2], the gate pulse amplitude $(V_{GH} - V_{GL})$ is kept constant and the charge pumping current is measured as a function of the base level, V_{GL}, of the pulse. A typical charge pumping current characteristic obtained in this manner is shown in Figure 3.10. As shown by Heremans et al [4], five distinct regions of operation can be identified. In regions 1 and 5 where the device is in accumulation during both ON and OFF periods or in inversion during both ON and OFF periods, no charge pumping current is measured.

Region 3 corresponds to the maximum CP current regime and extends over a range $V_T - \Delta V < V_{GL} < V_{FB}$ where ΔV is the amplitude of the pulse equal to $V_{GH} - V_{GL}$. Implicit in this condition is the assumption that the pulse amplitude ΔV is larger than $V_T - V_{FB}$. At $V_{GL} = V_T - \Delta V$, the surface starts to get inverted at the top of the pulse while the bottom of the pulse corresponds to accumulation. At $V_{GL} = V_{FB}$, the surface starts to get out of accumulation. The charge pumping current measured in region 3, is the

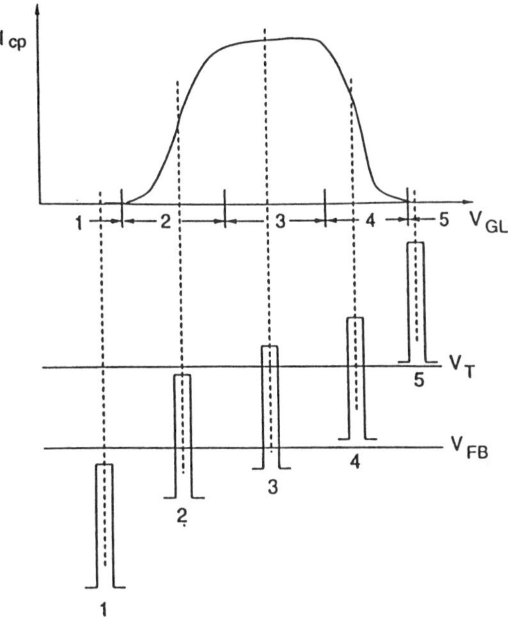

Figure 3.10: The CP current characteristics in the variable base mode. The pulse amplitude $V_{GH} - V_{GL}$ is kept fixed while V_{GL} is varied.

same as in region 3 in the variable amplitude technique (Figure 3.9) and equal to I_{CPmax}.

In region 2, the current rise with V_{GL} is similar to region 2 of the variable amplitude technique (Figure 3.9). Although V_{GL} is varying, the base of the pulse corresponds to accumulation. In region 2, the finite rising edge of the current characteristics is also due to the same reasons as discussed earlier in region 2 in the variable amplitude mode.

In region 4, the base of the pulse corresponds to a condition when the surface is slightly depleted. Nevertheless, we call this a weakly accumulated regime to show that there is a significant amount of majority carriers at the surface. During the ON period, the interface states get negatively charged by capturing electrons from the inversion layer at the surface. The finite slope of the falling edge can be due to non-uniformity in the flatband voltage V_{FB} i.e., a non-uniform channel. Even in an uniform channel, the finite slope can arise due the competing processes of electron emission and hole capture cited above.

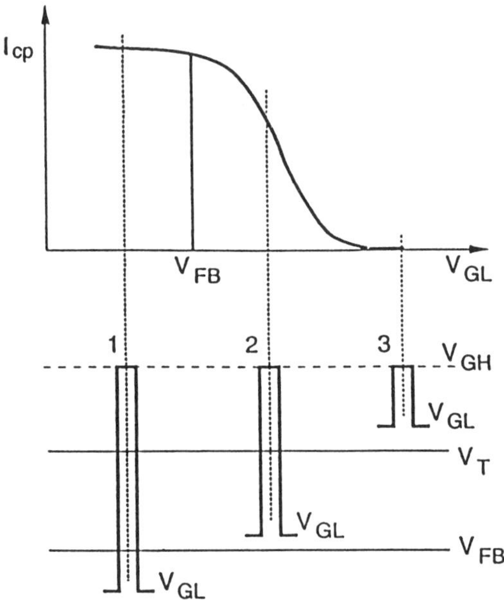

Figure 3.11: The CP current characteristics in the inverted variable amplitude mode.

3.4.3 The Inverted Variable Amplitude Mode

Still another variation of the technique is to use the variable amplitude technique with the top of the pulse V_{GH} fixed to correspond to strong inversion. A typical charge pumping current characteristic measured using this technique is shown in Figure 3.11. Again, three regions can be identified. In region 3, the surface is inverted in both ON and OFF period since V_{GL} is larger than V_T. Hence no charge pumping current is measured. In region 2, V_{GL} is at a level corresponding to weak accumulation.

The charge pumping current increases with a decrease in V_{GL} since more and more holes are available during the OFF period with decreasing V_{GL}. This region is similar to region 4 in variable base technique (Figure 3.10). In region 1, V_{GL} is sufficiently low ($V_{GL} < V_{FB}$) to keep the surface in accumulation during the OFF period. Hence, maximum charge pumping current is measured in this region similar to region 3 in both Figures 3.9 and 3.10. By analyzing the current characteristics in region 2, the non-uniformity of the flatband voltage in the channel can be studied.

3.4.4 Saw-tooth wave

There have been several other techniques used to measure the charge pumping current. One is to use a saw-tooth waveform rather than a rectangular pulse. The saw-tooth waveform is a limiting case of a trapezoidal pulse train (Figure 3.3) with T_{ON} and T_{OFF} becoming zero. Initially it was thought that the slow fall of the voltage during the decaying part of the saw-tooth will allow sufficient time for the mobile inversion carriers to flow back to the source and drain thereby eliminating or reducing the geometric component of the charge pumping current. This is not confirmed by experimental results. Increasing the rise time increases $E_{em,h}$ while increasing the fall time decreases $E_{em,e}$. The charge pumping current therefore decreases continuously with an increase in the rise and fall times of the pulse. The application of saw-tooth waveform is useful in obtaining more information on the interface emission parameters. A saw-tooth waveform with a short fall time and various rise times can be used to study hole emission parameters. Or alternately, use of a short rise time with various fall times of the saw tooth wave can be used to study electron emission characteristics of the interface states. The analysis is more elaborate but can be made with simplifying assumptions such as an uniform channel. As discussed later on, Groeseneken $et\ al$ [3] determined the geometric mean of capture cross sections of electrons and holes using saw-tooth waveforms of different rise and fall times.

3.4.5 Effect of Rise and Fall Times

The effect of the rise and fall times on the charge pumping current was modeled [3, 4] using a transient Shockley-Read-Hall (SRH) model developed by Simmons and Wei [6]. The charge pumping current was modeled as a function of the rise time T_R and fall time T_F and the interface state density profile was derived from the derivative of the charge pumping current versus T_R or T_F. The time taken for the gate voltage to change between V_{FB} and V_T is the time period $t_{em,h}$ equal to $t_2 + t_3$ in Figure 3.7 during which time period the hole emission takes place. During this period, holes will be emitted by the interface states having energy up to a level $E_{em,h}$, which is given by

$$E_i - E_{em,h} = kT\,ln\,[\,v_{th}\sigma_p n_i t_{em,h}\,] \qquad (3.22)$$

The interface states lying above $E_{em,h}$ would not have time to emit and become negatively charged by the time the gate voltage becomes equal to V_T and therefore are the ones available to capture inversion electrons. This level thus defines the lower level of the band of interface states that participates in charge pumping. Similarly, during the fall time of the gate pulse, the time interval $t_{em,e}$ in which the gate voltage sweeps from V_T to V_{FB} (equal to $t_2' + t_3'$ in Figure 3.7) determines the time period in which electrons are emitted. States above an energy level $E_{em,e}$ would have emitted the electrons where $E_{em,e}$ is given by

$$E_{em,e} - E_i = kT\,ln\,[\,v_{th}\sigma_n n_i t_{em,e}\,]. \qquad (3.23)$$

88

Only states lying below $E_{em,e}$ would not have emitted electrons by the time the gate voltage became equal to V_{FB} and thus be able to capture holes. Thus the level $E_{em,e}$ defines the upper level of the band of the interface states contributing to charge pumping. Hence the charge pumping current I_{CP} is equal to

$$I_{CP} = q\overline{D_{it}}Af\left[E_{em,e} - E_{em,h}\right] \tag{3.24}$$

$$= 2q\overline{D_{it}}AfkT\,ln\left[v_{th}n_i\sqrt{\sigma_n\sigma_p t_{em,e}t_{em,h}}\right]$$

where

$$t_{em,e} = \frac{V_T - V_{FB}}{V_{GH} - V_{GL}}T_F \tag{3.25}$$

$$t_{em,h} = \frac{V_T - V_{FB}}{V_{GH} - V_{GL}}T_R \tag{3.26}$$

and T_F and T_R are the fall and rise times of the square pulse. For a saw-tooth wave,

$$t_{em,e} = \frac{V_T - V_{FB}}{V_{GH} - V_{GL}}\frac{1}{f}\alpha \tag{3.27}$$

$$t_{em,h} = \frac{V_T - V_{FB}}{V_{GH} - V_{GL}}\frac{1}{f}(1 - \alpha) \tag{3.28}$$

where α is the fraction of the period when the gate voltage is rising and $(1 - \alpha)$ is the fraction of the period when the gate voltage is falling. The expression for I_{CP} should be accordingly modified for a saw-tooth wave as

$$I_{CP} = 2q\overline{D_{it}}AfkT\,ln\left[v_{th}n_i\sqrt{\sigma_n\sigma_p}\frac{|V_{FB} - V_T|}{(V_{GH} - V_{GL})}\frac{\sqrt{\alpha(1 - \alpha)}}{f}\right]. \tag{3.29}$$

By plotting the charge pumped per cycle (I_{CP}/f) as a function of f, a linear characteristic is obtained. From this plot, it is possible to obtain a geometric mean of the capture cross sections σ_n and σ_p in addition to $\overline{D_{it}}$. The above analysis assumes that the interface state density has an uniform distribution in the bandgap. In practice, the distribution is U-shaped and hence the plot of the charge per pulse as a function of emission time is not linear. The accuracy of the cross section measurement becomes unreliable. Another assumption is that during the transition from V_{FB} to V_T or from V_T to V_{FB}, the emission process dominates and that the rise and fall times are fast enough as not to let the interface attain equilibrium during the transition. The electron capture during the late stages of the transition from V_{FB} to V_T corresponding to weak inversion and hole capture during the late stages of the transition from V_T to V_{FB} corresponding to weak accumulation are neglected in this analysis. At lower temperatures, or for very short rise and fall times, the analysis becomes less applicable. The capture cross section is also assumed to be independent of energy in this analysis.

The analysis of the effect of rise and fall times on the CP current has been improved by Van den bosch *et al* [7] by including the hole emission during the period the pulse voltage changes from V_{GL} to V_{FB} (t_1 in Figure 3.7) during the rise time and the electron emission during the period the pulse voltage changes from V_{GH} to V_T (t_1' in Figure 3.7) during the fall time. They developed a technique to determine when the interface states reach a non-steady condition during the rise time. Time is measured from the instant the on-set of non-steady state condition occurs and the time during which non-steady state hole emission takes place is the time interval between $t = 0$ and the instant when the pulse voltage reaches a value V_T. From this, the energy level $E_{em,h}(T_R)$, up to which holes have been emitted during the rise time, is determined. Similarly the level down to which interface states have emitted electrons during the fall time, $E_{em,e}(T_F)$ is determined. The charge pumping action is due to interface states lying between $E_{em,e}(T_F)$ and $E_{em,h}(T_R)$. While this analysis is more detailed, the results approach the values given in Equations 3.24 and 3.29 when T_R and T_F are not too small. They used this analysis, to measure I_{CP} for 2 different values of T_{F1} and T_{F2} keeping T_R constant. The difference in I_{CP} between these two conditions is due to interface states lying between $E_{em,e}(T_{F1})$ and $E_{em,e}(T_{F2})$. Thus precise information is obtained regarding the interface states located in a narrow window of the band gap. By varying the temperature, the window can be moved in the band gap. The states in the upper half of the band gap are characterized using this method. Similarly keeping T_F constant and choosing two different rise times T_{R1} and T_{R2}, states in the lower half of the band gap can be studied using a temperature scan. This technique is called spectroscopic charge pumping.

3.4.6 Sensitivity of Charge Pumping Measurements

The charge pumping current is proportional to the pulse repetition frequency, and the number of interface states involved in the charge pumping process. The sensitivity of the dc current measuring instrument in the experimental set up determines the sensitivity of charge pumping technique in measuring the interface state density. Modern current meters can measure down to a few femtoamperes. However, measurement of such low currents requires a longer integration time and hence cannot be used at higher pulse repetition frequencies. It is possible to measure current levels of the order of a fraction of a picoampere with pulse repetition frequency of the order of 100 kHz. The charge pumped per pulse is $Q_{CP} = I_{CP}/f$. It is equal to $10^{-17}C$ (62 electrons) for a current of 1 pA at 100 kHz frequency. Assuming $W = 20\mu m$ and $L = 5\mu m$, the number of interface states measured is equal to $6.2 \times 10^7/cm^2$. If we assume $\Delta\Psi_S$ is 0.6 V between accumulation and inversion, an average surface state density $\overline{D_{it}}$ of $10^8/cm^2 eV$ can be measured. In modern scaled devices, typical values of the device dimensions are $W \approx 5\mu m$ and $L \approx 1\mu m$. Then the number of interface states that can be measured at this frequency and current level is equal to $\approx 1.2 \times 10^9/cm^2$. In scaled devices $\Delta\Psi_S$ is closer to 0.8 between accumulation and inversion. Therefore an average surface state density of $1.55 \times 10^9/cm^2 eV$ can be measured. We calculated these values assuming a value of 1 pA for the charge pumping current. But as stated earlier, it is possible to measure charge pumping current in the order of 0.1 pA. Hence a sensitivity of the order of $10^8/cm^2 eV$ can be realized even when using small geometry transistors. This approximate consideration of sensitivity in the charge pumping experiment illustrates that devices of the same size and geometry as those used in a VLSI circuit can be used in a charge pumping experiment.

The sensitivity of the charge pumping current measurement is affected by the presence of the geometric component. As stated before, this component arises due to the mobile inversion carriers that have not had time to flow back to the source and drain during the transition of the pulse from V_{GH} to V_{GL}. In those device structures where the geometric component is significant, a common practice is to use a large reverse voltage between the substrate and the source (and drain). The high electric field in the reverse-biased p-n junction acts as a sink for the charge carriers near the source and drain and therefore draws the carriers faster towards the source and drain thereby reducing the geometric component. Van den bosch *et al* [8] developed a method to reduce geometric component in the measured CP current. In this method, the interface state CP current and the geometric current are measured at different nodes. They use the well-substrate junction, which is reverse biased, as a collecting contact for most of the geometric current. When the pulse is OFF, the inversion carriers that are still left in the channel diffuse towards the well-substrate junction due to a strong vertical concentration gradient of the potential. This geometric component flows through the substrate contact. They used the technique to determine the presence of geometric component in a CP measurement. The current collected by the well contact is essentially the true interface CP current.

3.5 Applications of Charge Pumping

In this section, we will describe some specific applications of the CP technique.

3.5.1 Charge Pumping in Submicron Devices

The charge pumping technique is suitable for measurements on submicron devices. The sensitivity is sufficient to measure charge pumping current in small area devices. However, the assumption of an uniform channel condition is not valid in submicron devices. In short channel devices, the equipotential lines are curved and not parallel to the surface even in the middle of the channel. The bands bend more near the source and drain and less in the middle for a given gate voltage in depletion and inversion. Thus as the gate voltage is increased, inversion occurs first near the source and drain and spreads towards the middle with increasing gate voltage [9]. In Figure 3.12, the equipotential contours are shown for a gate voltage less than V_T. The equipotential line corresponding to $2\Phi_F$ intersects the surface at a point which is ΔL away from the source and at a point ΔL away from the drain. At these two points the band bending is

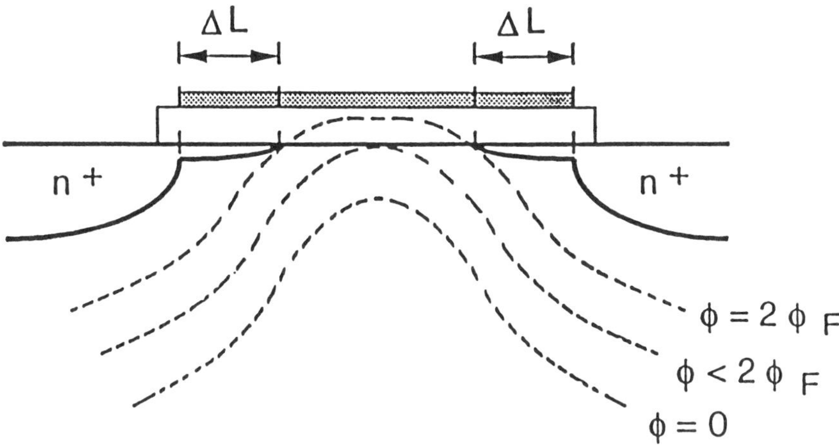

Figure 3.12: The equipotential contours in a short channel device. The contours are curved even in the middle of the channel.

$2\Phi_F$. Over the two regions of length ΔL there is inversion while in the middle region the band bending is less than $2\Phi_F$ and hence there is no inversion. The interface states distributed over the two regions of length ΔL will contribute to the charge pumping current. When V_{GH} exceeds V_T, the entire channel will be inverted and the interface states over the entire channel will be measured in the charge pumping experiment. If we further assume that in strong inversion the band bending Ψ_S is pinned at $2\Phi_F$, we can consider the channel as being uniform when $V_{GH} > V_T$. Using a two dimensional simulation technique, Gaitan *et al* [10] showed that the charge pumping model has decreasing accuracy as the effective channel length is reduced resulting in an over-estimation of the average interface state density. In short channel devices, the region of the channel influenced by the source and the drain becomes a larger fraction of the total channel leading to a variation in the threshold voltage throughout the channel. They showed that in short channel devices, the bell shaped CP current is skewed to one side and does not have a flat region.

Figure 3.13 shows the charge pumping current characteristics measured in a device with a channel length of 0.5 μm and a width of $10\mu m$. The oxide thickness is 90 Å. The results of variable amplitude measurement are given in Figure 3.13(a) while those of variable base measurement are given in Figure 3.13(b). The maximum charge pumping current corresponds to an average interface state density of $3 \times 10^{11}\ cm^{-2}eV^{-1}$. To get more quantitative values of interface state density distribution in energy as well as the profile of the interface state density along the channel, numerical simulations can be carried out [3]. The impurity profiles in the channel, source and drain are obtained by a process simulation program such as SUPREM III which is then converted into a two dimensional doping profile. Then a two dimensional simulator such as Pisces or Minimos is used to obtain the surface potential and electron concentration at the surface as a function of position in the channel for different values of V_{GH}. The interface states located at some point in the channel will contribute to charge pumping current if the top of the pulse V_{GH} is above the local threshold voltage and V_{GL} is below the local flat band voltage at that point. The total area of the channel which contributes to charge pumping equal to $2\Delta LW$ can be obtained for a specific pulse amplitude-base condition from simulation and the charge pumping current can be calculated. The simulation takes into account the recombination in weak inversion and weak accumulation. The agreement between measured and simulated characteristics suggests that charge pumping technique is valid for investigating the interface characteristics of short channel devices.

Hoffman *et al* [12] used a two dimensional simulator to calculate the charge pumping current in the time domain. They separate the surface charge into time dependent equilibrium and non-equilibrium components. The non-equilibrium component includes the transient effects due to the time constant effects of the interface states.

3.5.2 Charge Pumping at Low Temperatures

Till recently, charge pumping experiments have been confined mostly to temperatures above 77K. The difficulty in using charge pumping technique at lower temperatures arises because of substrate freeze-out effects. At very low temperatures, the substrate behaves like a leaky insulator and the flow of majority carriers to the surface during accumulation is limited by substrate time constant effects. Recently Nguyen-Duc *et al* [13] demonstrated for the first time that charge pumping measurements are feasible down to liquid helium temperature. They avoided the problem due to substrate freeze-out by having both electrons and holes needed for recombination travel along the surface. They used gate controlled MOS diodes in which the holes and electrons required for charge pumping action were supplied by the source and drain contacts of the opposite type located on the surface. For example if the source region was n^+, the drain region was p^+. They used both n^+np^+ or n^+pp^+ gate controlled diodes. The electrons flowed to the surface from the n^+ region during one phase of the pulse and holes flowed to the surface from the p^+ region during the opposite phase. At low temperatures, the time constant for the emission of holes and electrons is large and hence no emission takes place during rise and fall times in the charge pumping process for either carrier. The charge pumping action proceeds only by successive capture of electrons and holes from the surface. The channel length was chosen small enough to minimize geometrical component. They also studied the influence of the rise and fall times on the charge pumping current at 4.2, 77 and 300K. The Groeseneken analysis [3] to obtain the interface state energy profile from the derivative of the charge pump-

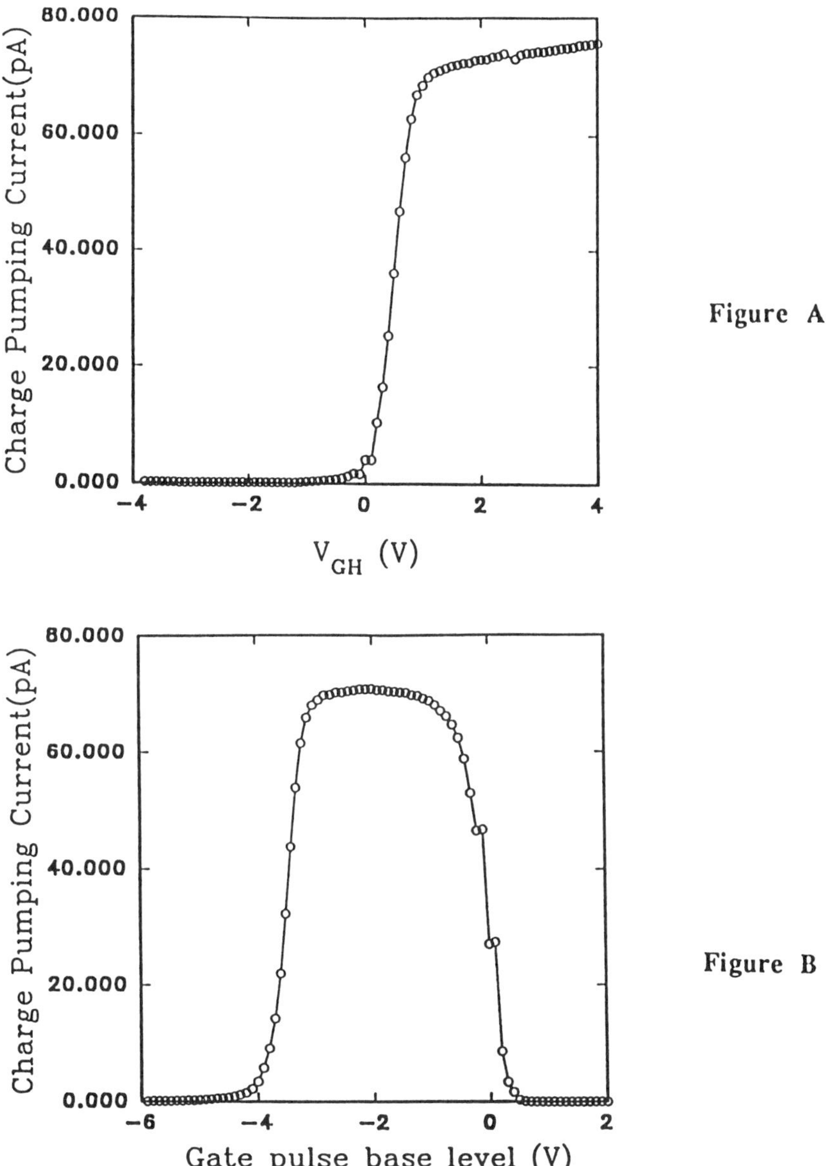

Figure 3.13: The plot of CP current in a short channel nMOS transistor in a CMOS wafer. The pulse frequency is 100 kHz. (a) Variable amplitude characteristic with the base voltage at -4V. (b) Variable base mode with a pulse amplitude of 4V.

ing current with respect to the rise or fall time was found to be valid only at higher temperatures. This is because at lower temperatures, the emission process is weak or almost non-existent whereas the Groeseneken analysis assumes that the emission process determines the energy band width within which the interface states contributing to the charge pumping current lie.

Hsu *et al* [14, 15] carried out charge pumping measurements on a small geometry MOS transistor at very low temperatures. In MOS transistors scaled down to small sizes, the substrate impurity concentration is made high as determined by an appropriate empirical scaling relationship. Increased substrate doping in scaled devices reduces the freeze-out effects. Hsu *et al* used MOS transistors with a substrate doping level of $\approx 4 \times 10^{17} cm^{-3}$ and $L = 0.5 \mu m$. Due to the high doping in the substrate, the freeze-out effect of impurities is not severe and hence charge pumping current can be measured. The current measured using the variable base technique at six different temperatures between 295K and 15K is shown in Figure 3.14. The average interface state density corresponding to the maximum charge pumping current measured at low temperature is roughly $5 \times 10^{10}/cm^2 eV$. The maximum charge pumping current is higher at lower temperatures since a larger region of the band gap is swept in going from accumulation to inversion and therefore more interface states contribute to the charge pumping current. Furthermore, at room temperature, the emission process takes away the contributions of some of the interface states as discussed earlier. But at lower temperatures, since the emission probability is decreased, the states which previously became negatively charged by hole emission now do so by capturing electrons and thereby contribute to charge pumping current. This is another reason why charge pumping current increases at lower temperatures. In a properly scaled device, the non-uniformity in V_T and V_{FB} is much less. The emission process also determines the finite slope of the rising and falling edges of the CP current. The rising and falling edges of the bell-shaped charge pumping current become steeper at lower temperatures due to the weakening of the emission process. The increasing edge is steeper because the electron density increases only when near threshold conditions are reached. Similarly the decreasing edge becomes steeper because the majority carrier density is negligible until V_{GH} exceeds the flatband voltage at which point the carrier density suddenly increases.

Low temperature charge pumping can be used to distinguish generation recombination (g-r) centers among interface states. g-r centers are those that have comparable capture cross-sections for both electrons and holes. On the other hand, traps can interact only with one type of carriers. An electron trap will capture electrons and become negatively charged during the ON period and become neutral during the OFF period by electron emission. A hole trap will similarly capture a hole during the OFF period and emit a hole during the ON period. Due to the decrease in emission probability with a decrease in temperature, the traps will contribute to charge pumping current at low temperature only if sufficient time is allowed for the emission to take place. If the pulse repetition frequency is varied (corresponding to varying the ON and

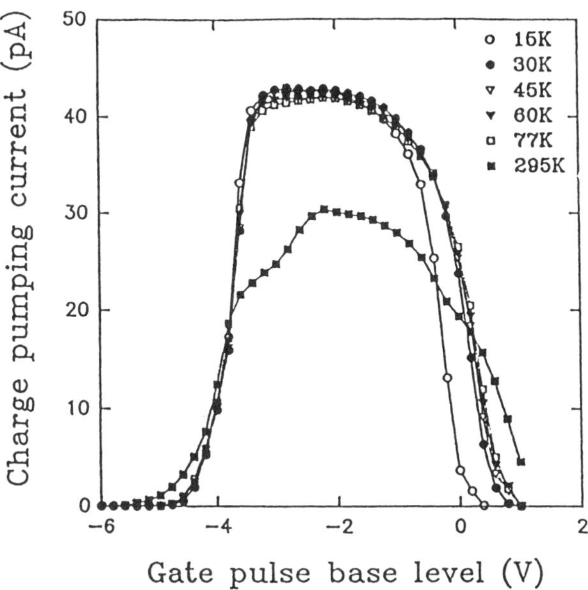

Figure 3.14: Plot of the CP current characteristics of an nMOS short channel transistor at various temperatures from 15K to 295K. The pulse amplitude was kept at 4V.

OFF periods) in a low temperature charge pumping experiment, the contribution of traps to the charge pumping current will decrease at higher frequencies while the contribution of the $g\text{-}r$ centers will remain frequency independent. In Figure 3.15, Q_{CP}, the charge pumped per pulse measured at 15K, is shown for different frequencies and no frequency dependence is seen. Hence it can be concluded that the interface states act only as $g\text{-}r$ centers and not as traps in this device.

3.5.3   Charge Pumping in Degraded Devices

The charge pumping technique has been employed to study the damage to the interface in devices exposed to ionizing radiation or subjected to electrical stresses such as Fowler-Nordheim ($F-N$) or hot carrier (HC) stress. Numerous authors have adapted the CP technique either directly or in a modified manner to investigate the effect of the stress.

The interface damage is more or less uniform over the entire channel length under exposure to ionizing radiation or under F-N stress. The damage is not uniform in the case of HC stress. Hot Carrier stress is produced in one of two ways. In one, called the Drain Avalanche Hot Carrier (DAHC) stress, electron-hole pairs are created due to the high electric field present in the drain depletion

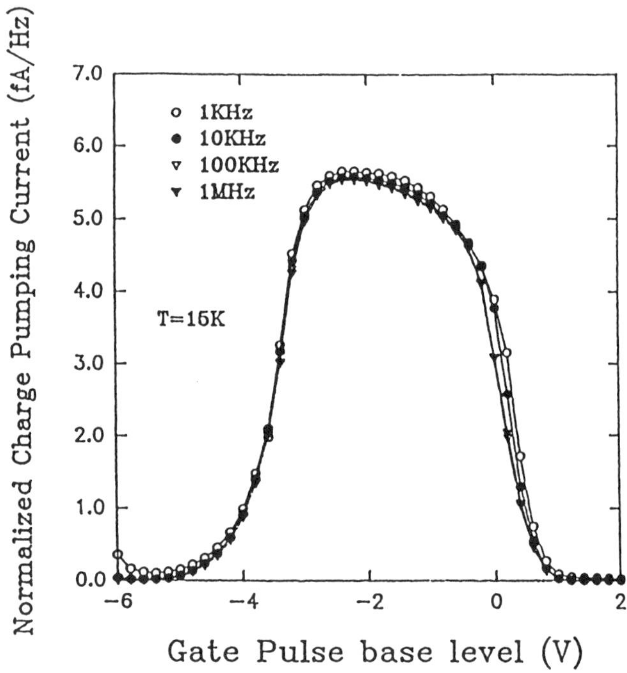

Figure 3.15: The plot of charge pumped per pulse in the variable base mode measured at 15K with the pulse repetition frequency varied between 1 kHz and 1 MHz in the same device as the one shown in Figure 3.15.

region and these hot carriers are injected into the oxide aided by the polarity of the electric field which is in such a direction as to drive the hot carriers towards the surface. At low gate voltages and for high drain voltages, the polarity of the electric field is such as to cause hot holes to be injected into the oxide.[2] At higher gate voltages approximately equal to half the drain voltage, the polarity is such that hot electrons are injected into the oxide. In either case, the hot carriers are injected into the oxide in a region close to the drain. This causes localized damage to the interface in the vicinity of the drain. It is believed that hot holes are injected closer to the drain than the hot electron injection. The non-uniformity in damage to the interface is again different between hot hole and hot electron injection.

The second mechanism of hot carrier injection is called the Channel Hot Carrier (CHC) stress or Channel Hot Electron (CHE) stress in our example of nMOS transistors. In this type of injection, channel inversion carriers are accelerated by the drain voltage and thus become hot. Some of these hot carriers get injected into the oxide. The region over which they are injected is larger

---

[2] Remember we are using nMOS transistors as an example in our discussion

in this case and thus the interface damage occurs over a larger region than in DAHC. Still, the damage is not uniform over the entire length of the channel.

The charge pumping technique is better suited to the study of non-uniform interface damage in MOS transistors than conventional techniques. As pointed out by Heremans *et al* [4], the channel can be treated as a composition of several regions each with a different value of threshold voltage. For example, the damaged region and the undamaged regions have different threshold voltages. The total charge pumping current is the sum of contributions from each of these regions. It is possible therefore to analyze the charge pumping current more directly or even design modifications of the measurement technique to investigate the interface state damage.

Heremans *et al* [16] determined the increase in interface state density after hot electron injection and after hot hole injection by measuring the difference in the maximum charge pumping current between virgin and stressed samples. The difference is only due to the contributions arising from the damaged region near the drain. Heremans *et al* [3] showed that the presence of fixed oxide charge due to the charge trapping as well as generation of interface states can be investigated using charge pumping techniques under all kinds of aging conditions such as exposure to radiation, F-N stress and HC stress. By dividing the channel into two regions, one damaged (near the drain) and the other undamaged (rest of the channel) and each with a different threshold voltage, they carried out a theoretical analysis of the charge pumping current and compared it with experimental results. The difference between the current measured during the increasing (rising) portion of the charge pumping characteristics with the drain connected and that measured under the same conditions with the drain floating is attributed to the contribution of interface states in the damaged region near the drain. In the range of voltages used in the measurements, the gate voltage is able to invert the region near the drain but not the rest of the channel. When the drain is connected, electrons are able to flow only from the drain (the source is not able to supply electrons because the rest of the channel is not inverted) and charge up the interface states giving rise to charge pumping current. When the drain is not connected, there is no way for electrons to come to this region and hence no charge pumping current is measured.

Vuillame *et al* [33] used simulations and charge pumping technique to study the creation of acceptor-like traps in the gate-drain overlap region of an nMOSFET subjected to hot electron and/or hot hole injection.

Using the method developed by Groeseneken *et al* [3], Chen *et al* [17] determined the interface state parameters in stressed samples. From the plot of the charge pumped per pulse as a function of frequency the geometric mean of the hole and electron capture cross sections of the interface states are obtained. They recorded the charge pumping current as the pulse frequency was swept over a range of two decades. This method assumes that the emission process is dominant and that the interface states are not in equilibrium during the transition from accumulation to inversion or from inversion to accumulation. Another implicit assumption is that the capture cross section is independent of energy and can be described by a single value for the interface states distributed over a range of energy.

3.5.4   Three Level Charge Pumping

Chung *et al* [18] used a three level method developed by Tseng [19] to measure interface state density. In Tseng's method, the gate pulse has in addition to the base and top levels, V_{GL} and V_{GH}, a third (intermediate) level V_e as shown in Figure 3.16. During T_1, all the interface states have captured holes and are in quasi-equilibrium with the holes in the accumulation region since V_{GL} is less than V_{FB}. During T_2, all the interface states have captured electrons and are in quasi-equilibrium.

When the gate pulse is switched from V_{GH} to the intermediate level V_e, the states emit electrons. If T_e is long enough, all states lying above an energy level E_{Te}, determined by V_e, will have emitted electrons and states below this level will still be negatively charged. Afterwards, during T_1, the interface states lying between E_{Te} and the Fermi energy level in accumulation will capture holes and thereby contribute to the charge pumping experiment. By measuring the charge pumping current as a function of V_e and therefore as a function of E_{Te}, the interface state density distribution is obtained. Chung *et al* [18] applied this technique to determine the emission time constant characteristics of the interface states by shortening T_e. They applied this technique to determine the increase in interface state density due to F-N stress and hot electron stress. Saks *et al* [20] extended the three level technique to measure the electron and hole capture cross sections of the interface states. The energy level of the interface states up to which emission takes place is a function of V_e if T_e is long enough. It will be a function of T_e if the emission process is not completed at the end of T_e. In the plot of the charge pumping current as a function of T_e, the value of T_e at which the current levels off is taken as the emission time constant. From a knowledge of the emission time constant and the energy of the interface state, which is obtained from V_e, the capture cross section is obtained.

Cilingiroglu [21] modified the CP technique by combining it with the pulsed interface probing (PIP) technique [22]. Two different pulse waveforms are used in this technique. One is the usual rectangular (or trapezoidal including the rise and fall times) which pulses the device between accumulation (during the OFF period) and depletion (during the ON period). The other waveform has a narrow pulse of duration T_I superimposed in the middle of the depletion phase as shown in Figure 3.17. This narrow pulse inverts the surface to charge the interface states negatively. During the ON period, the time interval between the start of the ON period and the start of the narrow pulse, denoted T_L, determines the lower level of the energy range of the contributing interface states. Similarly the time interval between the end of the narrow pulse and the end of the ON period, denoted T_U determines the upper level of the energy range of the contributing interface states under the assumption that T_L and T_U are not long enough for the interface states to reach equilibrium. The

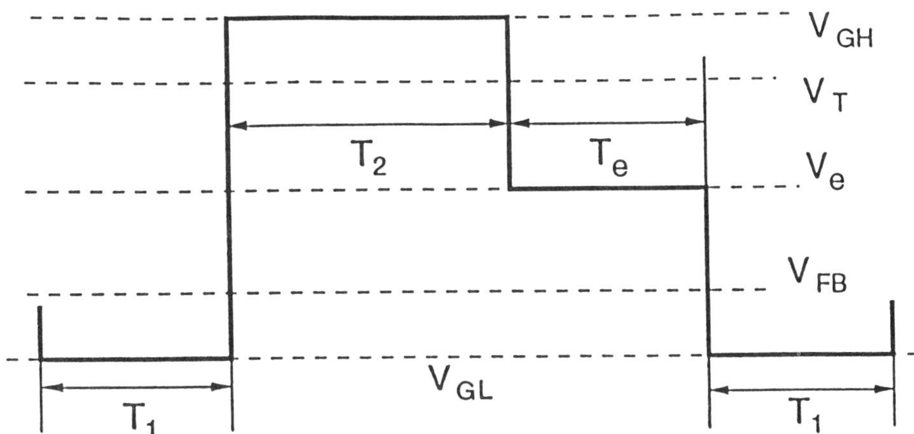

Figure 3.16: Three level gate pulse: In addition to V_{GL} and V_{GH} the pulse has an intermediate level V_e.

source current is measured using successively each of the two waveforms. The difference in current between the two waveforms is taken as a direct measure of the CP current due to the interface states and the subtraction procedure eliminates the leakage current in the junction. The interface state density in the lower half of the bandgap is obtained by measuring the source CP current, I_{CPS} as a function of T_L and similarly by varying T_U the interface state density in the upper half of the bandgap is obtained. In another modification of the CP technique, Kejhar [23] replaces the reverse bias source/drain supply voltage by a second pulse generator which is synchronized to the gate pulse. This technique is called the Double-Pulse charge pumping. The major advantage of this technique over the three level charge pumping is claimed to be enhanced immunity against the parasitic geometric component.

3.5.5 Lateral Profiling of Interface States

Ancona *et al* [24] measured the spatial variation of the interface states using a technique called "constant field" profiling technique. By varying the source/drain reverse voltage the effective channel length is varied. The variation in the current I_{CP} gives information on the interface state density profile. However, when the source/drain voltage is varied the threshold voltage is also varied. To obviate this difficulty Ancona *et al* [24] applied the source/drain bias voltage in the form of a pulse which is on only during the accumulation phase of the gate voltage pulse and not during the inversion phase. Thus the problem of threshold voltage variation with the source/drain reverse voltage is eliminated.

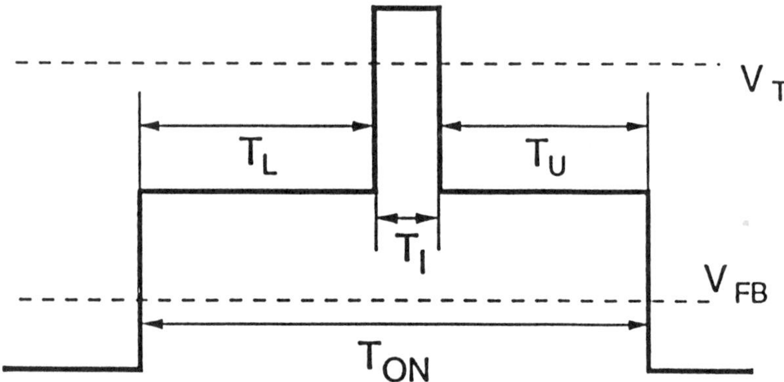

Figure 3.17: The pulse on pulse waveform used by Cilingirogulu [21]. The narrow pulse inverts the surface while during the rest of the ON period the channel is depleted.

Chen *et al* [25, 26] developed a new technique to measure the interface characteristics in non-uniform channels as in stressed devices. During the rising edge of the charge pumping characteristics, a small increment in V_{GH} causes an elementary increase in the current. This current is due to a small increase in the inverted portion of the channel. The gate voltage V_{GH} at which this increment occurs is the local threshold voltage in this elementary portion of the channel. The derivative of charge pumping current will be non-zero only if V_{GH} is equal to the threshold voltage and is given by

$$\frac{dI_{CP}(V_{GH})}{dV_{GH}} = qWf \int_0^L N_{it}(y)\, \delta\left[V_{GH} - V_T(y)\right]\, dy \qquad (3.30)$$

where I_{CP} is the charge pumping current, W and L are the width and length of the channel, f the pulse frequency, δ the Dirac delta function, $N_{it}(y)$ the distribution of interface states along the channel integrated between $E_{em,h}$ and $E_{em,e}$. $V_T(y)$ is the local threshold voltage at y. The contribution to $dI_{CP}(V_{GH})/dV_{GH}$ comes from those regions where V_{GH} is equal to the local threshold voltage V_T. If the threshold voltage is uniform (*i.e.* the same everywhere), the derivative of I_{CP} will be sharply peaked. If the channel is non-uniform the plot of the derivative of I_{CP} will exhibit a finite width representing the distribution of local threshold voltages $V_T(y)$ in the channel.

The lateral non-uniformity in the threshold voltage is caused primarily by localized charges near the interface. As discussed previously, if the reverse-bias between the substrate and the source/drain junctions is varied, the channel length is varied. For an increase ΔV_R in the reverse bias, the channel length L is decreased by ΔL with a corresponding decrease in I_{CP}. The change in the maximum value of I_{CP} is given by

$$\Delta I_{CPmax} = qWfN_{it}(L')\Delta L \qquad (3.31)$$

where $L' = L - \Delta L$. Using a simulation program, one can determine the variation in channel length with the reverse voltage V_R. In the experiment, a variable reverse voltage is applied on the source/drain junction and a series of I_{CP} vs V_{GH} and dI_{CP}/dV_{GH} vs V_{GH} curves are obtained for different reverse junction voltages. From the incremental change of I_{CPmax}, $N_{it}(y)$ is obtained and the voltage where the incremental change occurs gives the threshold voltage $V_T(y)$ from which the trapped charge $N_{it}(y)$ can be obtained. The constant field approach of Ancona et al [24] was adapted by Chen et al and the drain voltage was pulsed.

Li et al [27] used a spatial profiling technique in which the source and the drain regions are independently reverse-biased. By varying the values of the source reverse bias keeping the drain reverse bias the spatial variation in the interface state density near the source is obtained. Doing the reverse, the profile near the drain is obtained.

Another new CP method has been developed by Tsuchiaki et al [28] and applied to study stressed devices. This technique is particularly suitable for the investigation of non-uniform interfaces and determines the distribution of fixed charge near the drain junction. In nMOSFETs, the interface state generation is a major degradation mechanism and the charge pumping technique is suitable for measuring interface state density. In pMOSFET devices electron trapping is the dominant degradation mechanism. This technique is well suited for investigating the lateral distribution of fixed charge in stressed pMOSFET devices.

Assuming a variation in the local threshold voltage $V_T(y)$ and $V_{FB}(y)$ in the channel, the time for emission of electrons during the switching of the pulse from accumulation to inversion denoted $t_{em,e}(y)$ and the time for emission of holes during the transition from inversion to accumulation denoted $t_{em,h}(y)$ determine the contribution of interface states at y to the CP current. As V_{GH} is incremented by a small amount, the corresponding increase ΔI_{CP} gives the contribution from an elementary section located between y and $y + dy$, to the charge pumping current. The elementary section dy has a local threshold voltage equal to V_{GH}. Instead of keeping the rise time and fall time constant when V_{GH} and/or V_{GL} are varied, the gradient of the rise and fall time is kept constant. This assures that the contribution of each elementary section is independent of V_{GH} and V_{GL}. The source and the drain CP current, I_{CPS} and I_{CPD} are measured (Figure 3.5) and the range over which I_{CPS} or I_{CPD} increases from 0 to a saturation value yields the range of local threshold voltage variation in the channel. Combining this with a two dimensional program the distribution of fixed charges in the oxide can be obtained.

3.5.6   CP Technique Applied to SOI devices

A number of workers [40, 30, 31] have successfully applied the CP technique to study silicon-on-insulator (SOI) MOS transistors. The SOI device performance is determined by both the front and back interfaces. The devices are leaky and unstable in comparison with the bulk devices and hence the typical capacitance measurement techniques are not successful. The CP technique is more successful. The front interface is studied by pulsing the gate while the back interface is studied by pulsing the substrate. Wouters *et al* [40] showed, using the CP technique, that the front interface in a laser-recrystallized SOI MOSFET is almost as good as the interface in a bulk device while the back interface is of lower quality. Ouisse *et al* [30] measured CP current in thin film SOI MOSFET as a function of the voltage on the back gate with different rise and fall times for the gate pulse. They found evidence of geometric component at high back gate bias values. They also studied the influence of channel length on CP current. Seghir *et al* [31] compared the leakage current measurement technique and the CP current measurement using a gate controlled diode structure on SOI material and concluded that the two techniques yield complementary results and that both techniques are useful.

3.5.7   Slow State Investigation

Paulsen *et al* [32] extended the charge pumping technique to study slow states *i.e.* those located in the oxide within a tunneling distance from the Si-SiO$_2$ interface. At the frequencies used in usual CP measurements, Q_{CP}, the charge pumped per pulse is independent of frequency, and is due to the charging and discharging of fast states. The slow states are not able to follow the rapid switching of the surface between accumulation and inversion. However, if the pulse repetition frequency is reduced considerably, allowing the time for carriers to tunnel in and out of the slow states, the charge pumping current will then include the contribution of slow states also. When Q_{CP} is plotted as a function of frequency, the value increases with decreasing frequency below a breakpoint while it is a constant with frequency above the breakpoint. The reciprocal of this frequency is taken as the tunneling time constant. Using a simplified tunneling model, the tunneling distance can be calculated. This technique has been used by Paulsen *et al* [32] to investigate radiation damaged interfaces where they observe an increase in both fast and slow interface states.

3.6   Conclusion

The discussion in this chapter illustrates that the CP technique is a powerful method to investigate the Si $-$ SiO$_2$ interface in MOS transistors. The progress in the field of microelectronics has made possible the fabrication and use of MOS transistors with channel length in the sub-micron range. The CP technique is able to meet the challenges posed in the characterization of submicron transistors and yield accurate results. The understanding of the electronic processes in a charge pumping experiment has improved with time. This makes it possible to investigate a wide range of parameters in short channel devices. The need for a specially designed test device structure is avoided since CP measurements can be directly made on the transistor whose interface characterization is needed. The CP technique can be used for a simple and rapid measurement of the average interface state density from I_{CPmax}, the maximum CP current. It can be used to monitor the process as devices come off the line. CP measurements can be carried out on a wafer and therefore special device mountings are not needed. When accurate and detailed measurements are needed to study the interface characteristics, various sophisticated modifications of the basic CP technique, that have been developed in the last ten years, can be used.

CP technique lends itself nicely to investigate non-uniform degradation in MOS transistors. Thus it is a valuable tool in the study of hot carrier lifetime and reliability. CP technique is also useful in the study of radiation damage and F-N stressed devices. The technique is also being used to study process-induced damage in sub-micron devices.

The important experimental control parameters in charge pumping are V_{GL}, V_{GH}, T_{ON}, T_{OFF}, T_R and T_F. By varying these parameters, detailed information regarding the interface states is obtained. By varying the temperature, additional information is also obtained. CP technique can be used on scaled devices at cryogenic temperatures. Low temperature measurement enables the scanning of the region of the band gap closer to the band edges.

The value of I_{CPmax}, the rising and falling edge of I_{CP} characteristics and their dependence on rise and fall times give complete information on the interface conditions in the device. The rising and falling edges of I_{CP} characteristics yield information on channel uniformity. While CP measurements are normally used to measure fast interface state characteristics, CP technique is increasingly used for the profiling of slow states. CP technique has proved to be a versatile measurement technique warranting its use for process control, device analysis and characterization, reliability, and stress studies.

Bibliography

[1] J. Stephen Burgler and Paul G. Jespers, "Charge Pumping in MOS Transistors", IEEE Trans. El. Dev., ED-16, p. 297, 1969.

[2] Alexander B.M. Elliot, "The Use of Charge Pumping Currents to Measure Surface State Densities in MOS Transistors", Sol. St. El., 19, p. 241, 1976.

[3] Guido Groeseneken, Herman E. Maes, Nicolas Beltran, and Roger F. De Keersmaecker, "A Reliable Approach to Charge-Pumping Measurements in MOS Transistors", IEEE Trans. El. Dev., ED-31, p. 42, 1984.

[4] Paul Heremans, Johan Witters, Guido Groeseneken and Herman Maes, "Analysis of the Charge Pumping Technique and its Application for the Evaluation of MOSFET Degradation", IEEE Trans. El. Dev., ED-36, p. 1318, 1989.

[5] W.V. Backensto and C.R. Viswanathan, "Measurement of Interface State Characteristics of MOS Transistor Utilizing Charge Pumping Techniques", Proc. IEE, Vol 128, pt. 1, p. 44, 1981.

[6] J.G. Simmons and L.S. Wei, "Theory of Dynamic Charge Current and Capacitance Characteristisc in MIS Systems Containing Distributed Surface Traps", Sol. St. El., Vol 16, p. 53, 1973.

[7] Geert Van den bosch, Guido V. Groeseneken, and Herman Maes, "Spectroscopic Charge Pumping: A New Procedure for Measuring Interface Trap Distributions on MOS Transistors", IEEE Trans. El. Dev., Vol 38, p. 1820, 1991.

[8] Geert Van den bosch, Guido V. Groesenekan, and Herman Maes, "On the Geometric Component of Charge-Pumping Current in MOSFET's", IEEE El. Dev. Lettr., Vol 14, p. 107, 1993.

[9] C.R. Viswanathan, B.C. Burkey, G. Lubberts, and T.J. Tredwell, "Threshold Voltage in Short-Channel MOS Devices", IEEE Trans. El. Dev., ED-32, p. 932, 1985.

106

[10] M. Gaitan, E.W. Enlow, and T.J. Russell, "Accuracy of the Charge Pumping Technique for Small Geometry MOSFET's", IEEE Trans. Nucl. Sci., Vol 36, p. 1990, 1989.

[11] Jen-Tai Hsu, "The Effect of Hot-Carrier and Fowler-Nordheim Injection on VLSI MOSFET at Room and Cryogenic Temperatures", Ph.D. Dissertation, UCLA, 1993.

[12] F. Hofmann and W. Hansch, "The Charge Pumping Method: Experiment and Complete Simulation", J. Apply. Phys., Vol 66, p. 3092, 1989.

[13] CH. Nguyen-Duc, G. Ghibaudo, and F. Balestra, "Validation of the Charge Pumping Method Down to Liquid Helium Temperature", Phys. Stat. Sol. (a), 126, p. k139, 1991.

[14] Jen-Tai Hsu, C.R. Viswanathan, R. Divakaruni, and Xiaoyu Li, "Charge Pumping at Cryogenic Temperatures", Proceedings of the 23rd European Solid State Device Research Conference, Grenoble, France, pp. 527-530, 1993.

[15] Jen-Tai Hsu and C.R. Viswanathan, "Identification of Generation-Recombination Centers and Traps Among Interface States by Cryogenic Charge Pumping Technique", Japan. Journ. Appl. Phys., Vol 33, p. 247, 1994.

[16] P. Heremans, H.E. Maes, and N. Saks, "Evaluation of Hot Carrier Degradation of N-Channel MOSFET's with the Charge Pumping Technique", IEEE El. Dev. Lett., EDL-7, p. 428, 1986.

[17] Wenliang Chen, Artur Balasinski, Binglong Zhang, and Tso-Ping Ma, "Hot Carrier Effects on Interface-Trap Capture Cross Sections in MOSFET's as Studied by Charge Pumping", IEEE El. Dev. Lettr., EDL-13, 1992.

[18] James E. Chung and Richard S. Muller, "The Development and Application of a $Si - SiO_2$ Interface-Trap Measurement System Based on The Staircase Charge-Pumping Technique", Solid St. Electron., Vol 32, p. 867, 1989.

[19] W.L. Tseng, "A New Charge Pumping Method of Measuring $Si - SiO_2$ Interface States", Journ. Appl. Phys., Vol 62, p. 591, 1987.

[20] Nelson S. Saks and Mario G. Ancona, "Determination of Interface Trap Capture Cross Section Using Three-Level Charge Pumping", IEEE El. Dev. Lettr., EDL-11, p. 339, 1990.

[21] Ugur Cilingiroglu, "Charge Pumping Spectroscopy with Pulsed Interface Probing", IEEE Trans. El. Dev., ED-37, p. 267, 1990.

[22] Ugur Cilingiroglu, "A Pulsed Interface-Probing Technique for MOS Interface Characterization at Midgap Levels", IEEE Trans. El. Dev., ED-35, p. 2391, 1988.

[23] Martin Kejhar, "Double-Pulse Charge Pumping in MOSFET's", IEEE El. Dev. Lettr., EDL-6, p. 344, 1992.

[24] M.G. Ancona, N.S. Saks, and D. McCarthy, "Lateral Distribution of Hot Carrier-Induced Interface Traps in MOSFET's", IEEE Trans. El. Dev., ED-35, p. 2221, 1988.

[25] Wenliang Chen, and Tso-Ping Ma, "A New Technique for Measuring Lateral Distribution of Oxide Charge and Interface Traps Near MOSFET Junctions", IEEE El. Dev. Lettr., EDL-12, p. 393, 1991.

[26] Wenliang Chen, Artur Balasinski, and Tso-Ping Ma, "Lateral Profiling of Oxide Charge and Interface Traps Near MOSFET Junctions", IEEE Trans. El. Dev., ED-40, p. 187, 1993.

[27] X.M. Li and M.J. Deen, "Determination of Interface State Density in MOSFET's Using the Spatial Profiling Charge Pumping Technique", Sol. St. Electron., Vol 35, p. 1059, 1992.

[28] Masakatsu Tsuchiaki, Hisahi Hara, Toyota Morimoto, and Hiroshi Iwai, "A New Charge Pumping Method for Determining the Spatial Distribution of Hot-Carrier-Induced Fixed Charge in p-MOSFET's", IEEE Trans. El. Dev., ED-40, p. 1768, 1993.

[29] Dirk J. Wouters, Marnix R. Tack, Guido Groeseneken, Herman E. Maes, and Cor L. Claeys, "Characterization of Front and Back $Si - SiO_2$ Interfaces in Thick and Thin Film Silicon-on-Insulator MOS Structures by the Charge Pumping Technique", IEEE Trans. El. Dev., Vol 36, p. 1746, 1989.

[30] Thierry Ouisse, Sorin Cristoloveanu, Tarek Elewa, Hisham Haddara, Gerard Bonel, and Dimitris E. Ioannou, "Adaptation of the Charge Pumping Technique to Gated $p - i - n$ Diodes Fabricated on Silicon on Insulator", IEEE Trans. El. Dev., Vol 38, p. 1432, 1991.

[31] Hachemi Seghir, Sorin Cristoloveanu, Robert Jerisian, Jean Ovalid, and Andre-Jacques Auberton-Herve, "Correlation of the Leakage Current and Charge Pumping in Silicon on Insulator Gate-Controlled Diodes", IEEE Trans. El. Dev., Vol 40, p. 1104, 1993.

[32] R. E. Paulsen, R. R. Siergiej, M. L. French, and M. H. White, "Observation of Near-Interface Oxide Traps with the Charge-Pumping Technique", IEEE El. Dev. Lettr., Vol 13, p. 627, 1992.

[33] Dominique Vuillame, Jean Claude Marchetaux, and Alain Boudou, "Evidence of Acceptor-Like Oxide Defects Created By Hot-Carrier Injection in n-MOSFET's : A Charge Pumping Study", IEEE El. Dev. Lett., EDL-12, p. 60, 1991.

[34] Artur Balasinski and Tso-Ping Ma, "Enhanced Electron Trapping Near Channel Edges in NMOS Transistors", IEEE Trans. El. Dev., ED-39, p. 1680, 1992.

4

Deep Level Transient Spectroscopy (DLTS)

Peter McLarty
North Carolina State University
North Carolina, USA

4.1 Introduction

Many semiconductor defects manifest themselves through deep and shallow levels in the energy gap. Deep levels may act as electron or hole traps depending on their state of occupancy. Shallow levels are formed in semiconductor crystals when a foreign atom, which belongs to the groups of the periodic table closest to that of the semiconductor, is introduced into the lattice. These shallow levels are used to control the concentartion of holes or electrons in the semiconductor. They can be described by the effective mass theory of Kohn and Luttinger [1], giving a hydrogen like spectrum of the discrete levels with the binding energies E_n, which can be written as:

$$E_n = \frac{1}{n^2} \frac{m^*}{2h} \left(\frac{q^2}{\epsilon} \right)^2 \tag{4.1}$$

where m^* is an effective mass for the electron or hole, ϵ is the static dielectric constant of the semiconductor, and n is an integer. For the ground state energies, better agreement between calculated and measured values are obtained using a more complicated pseudo-impurity theory developed by Pantelides [2].

Impurity levels which lie at a considerable distance ($> 0.2\ eV$) from a band edge are called deep level impurity states. The large ionization energy of these states implies a strong potential which acts to localize the carrier waveform near the site of the defect. One important consequence of this localization is the delocatization in k-space which results in strong lattice coupling and the consequent non-radiative nature of these deep states. Because of this non-radiative nature, luminescence measurements do not provide promising methods for their study.

A number of electrical methods for the study and characterization of deep levels in semiconductors have been developed in the past decade and are discussed in several excellent review articles. Lang [3], Schroder [4], and Sah [5], among others, review the application of space charge regions to the study of deep levels. Miller *et al* [6], discuss capacitance transient methods developed to measure thermal emission and capture rates. Currently, the most widely used measurement for the study of deep levels in semiconductors is Deep Level Transient Spectroscopy (DLTS) introduced by Lang [7]. In its original form, DLTS is a capacitance transient thermal scanning technique used for measuring the capture cross-section, concentration, and activation energies of deep levels. The main advantages of DLTS over previous capacitance transient techniques are the relatively short measurement times and ease of implementation. It has been applied using the capacitance of pn junctions, Schottky diodes, and MOS capacitors. In certain instances the current in these structures is monitored instead of the capacitance [8], [9], [10]. To investigate the defects existing both in thin films and processed devices, the DLTS concept has been applied using FET transistors [11]. These transistors include both enhancement and depletion mode MOSFETs and MESFETs [12], [13], [14].

As the device size is reduced, the capacitances involved become very small, making the measurement of the small capacitance changes required in DLTS very difficult. To alliviate this limitation, current DLTS has replaced conventional DLTS in making measurements on small devices. In this chapter the discussion will predominantly focus on the application of the current DLTS technique using both enhancement and depletion mode MOSFETs .

4.2   Generation, Recombination and Trapping Statistics

The MOSFET current DLTS technique involves the controlled filling and emptying of deep levels (traps) and the monitoring of the resulting drain current transients. These transients can result from traps located in the semiconductor substrate, the semiconductor-silicon dioxide interface, and in the bulk oxide region close to the semiconductor oxide interface. The statistics which describe

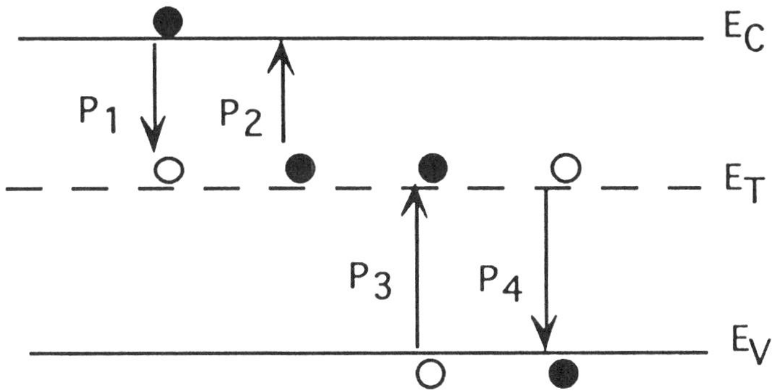

Figure 4.1: Energy band diagram illustrating the various capture and emission processes from a deep level

these processes will be reviewed to provide a suitable background for the discussion of the MOSFET current DLTS technique.

4.2.1 Shockley-Hall-Read Statistics

Consider a deep state with concentration N_T with an energy level E_T below the conduction band edge. The four possible processes which can take place from this state are illustrated in Figure 4.1 and consist of:

- electron capture from the conduction band (P_1)

- electron emission to the conduction band (P_2)

- hole capture from the valence band (P_3) and,

- hole emission to the valence band (P_4).

The process of electron capture can be understood if one considers the physical processes involved. Electron capture represents the movement of an electron from the conduction band to the deep state. For this to happen two conditions must be satisfied. There must be electrons in the conduction band to make the transition and there must be unoccupied deep states to recieve the electrons. The process of electron capture is proportional to the electron concentration n and the number of empty states given by $N_T[1 - f(E_T)]$. Here $f(E_T)$ the Boltzmann distribution function which describes the occupation probability of the state. The constant of proportionality for the capture process is the probability per unit time that an electron will make the transition. This leads to the following expression for the electron capture process P_1

$$P_1 = v_{th}\sigma_n n N_T[1 - f(E_T)] \tag{4.2}$$

112

where σ_n is the capture cross section and v_{th} the thermal velocity. The Boltzmann approximation $f(E_T)$ is valid as long as the deep state is more than 3kT from the band edge. From Equation 4.2, the rate of carrier capture c_n to the deep state from the conduction band can be written as:

$$c_n = v_{th}\sigma_n n \qquad (4.3)$$

For electron emission to the conduction band the only necessary criteria is the presence of the deep states occupied with electrons. The resulting expression for the electron emission to the conduction band is as follows:

$$P_2 = N_T f(E_T) e_n \qquad (4.4)$$

Here e_n is the probability that the electron will make the transition to the conduction band. In thermal equilibrium the processes P_1 and P_2 are equal which results in the following expression for the rate of electron emission e_n :

$$e_n = v_{th}\sigma_n n_i \exp\left(-\frac{E_c - E_T}{kT}\right) \qquad (4.5)$$

Hole capture involves the movement of a hole from the valence band to the state. This process is equivalent to an electron leaving the state and entering the valence band. Using arguments very similar to those presented above, the expression for hole capture is given by:

$$P_3 = v_{th}\sigma_p p f(E_T) N_T \qquad (4.6)$$

and that for hole emission by:

$$P_4 = N_T[1 - f(E_T)]e_p \qquad (4.7)$$

From Equation 4.6, the hole capture rate c_p can be written as:

$$c_p = v_{th}\sigma_p p \qquad (4.8)$$

and using thermal equilibrium arguments, the rate of hole emission is given by:

$$e_p = v_{th}\sigma_p n_i \exp\left(-\frac{E_T - E_v}{kT}\right) \qquad (4.9)$$

The classification of the deep state as a generation center, recombination center, or trap depends on the order in which the various processes take place. To illustrate this concept, start by considering electrons in the conduction band. Some of these electrons will be captured by the deep states, a process described by P_1. Once in the deep state there are two possible actions:

1. the electrons can be re-emitted to the conduction band, P_2, or

2. hole capture can take place, P_3, which results in the movement of the electron from the deep state to the valence band.

If P_1 follows P_2 as described in (1) then the deep state is said to be acting like a trap. If P_3 follows P_1 however, then the state is a generation/recombination center. The same classification applies for hole capture from the valence band, i.e. $P_3 - P_4$ describes a hole trap while $P_3 - P_1$ a generation/recombination center. The factors which determine which process dominates are the location of the Fermi level, the capture cross-section of the deep state and the temperature. In general however, the states close to mid-gap tend to behave as generation/recombination centers while those close to the band edges behave as traps. States located in the upper half of the band gap are electron traps and have electron emission rates that are much larger than their hole emission rates. States in the lower half of the band gap behave as hole traps with a correspondingly larger hole emission rate. The different emission rates for traps located in the various regions of the band gap are used in MOSFET current transient spectroscopy to simplyfy the analysis.

Now consider the specific case of a p-type semiconductor with an acceptor trap at a level E_a above the valence band edge and develop a set of equations that describe the occupation of the level. Let the number of traps occupied with holes be p_T, those occupied with electrons n_T, and the total number of traps N_T. It then follows that the total number of traps must be equal to the total number occupied by electrons plus the total number occupied by holes or, $N_T = p_T + n_T$. The rate of change of hole concentration at the deep state dp_T/dt can be written as

$$\frac{dp_T}{dt} = e_n n_T + c_p p n_T - e_p p_T \tag{4.10}$$

This equation can be easily solved and results in the following expression for $p_T(t)$

$$p_T(t) = p_{Ti} \exp\left(\frac{-t}{\tau}\right) + \frac{N_T(e_n + c_p p)}{e_n + c_p p + e_p}\left[1 - \exp\left(\frac{-t}{\tau}\right)\right] \tag{4.11}$$

Here τ is given by $1/(e_n + c_p p + e_p)$. It has been assumed that at $t = 0$ the hole trap concentration $p_T = p_{Ti}$. This solution can be further simplified if one considers two special cases which are both easily obtained in practice. If all the traps in a depletion region with concentration N_T are initially filled with holes then all the rates are set to zero except e_p. Equation 4.11 then reduces to the following:

$$p(t) = N_T \exp(-e_p t) \tag{4.12}$$

If on the other hand one considers an initially empty trap in the quasineutral region the solution becomes

$$p(t) = N_T \left[1 - \exp(-c_p p t)\right] \tag{4.13}$$

Equations 4.12 and 4.13 will be used later in developing the equations that describe the MOSFET current DLTS technique.

4.3 MOSFET Current Transient Spectroscopy

Enhancement MOSFET current transient spectroscopy was first applied to study the properties of deep levels existing in thin silicon-on-sapphire films [15]. The very nature of these films and the high series resistances involved precluded the use of capacitance DLTS measurements. To overcome the series resistance and interface coupling problems in the study of deep states existing in these thin SOI films, current DLTS was applied using depletion MOSFET's [13] and enhancement MOSFET's [16]. While both of the above techniques can give information concerning the deep traps in the material, the signal obtained using depletion mode transistors is influenced to a large degree by states existing at the gate SiO_2/Si interface. The presence of the inversion layer in enhancement-mode transistors at this interface effectively screens these states during the measurement. For this reason enhancement mode MOSFET current transient spectroscopy will be discussed in detail in the following sections with a brief qualitative description of the depletion mode technique included for completeness.

4.3.1 Enhancement MOSFET Current DLTS

First consider the experimental procedure used in enhancement MOSFET current deep level transient spectroscopy. In MOSFET current DLTS an enhancement mode MOSFET is repeatedly switched from accumulation to inversion and the effect of trap emission on the drain current monitored. Here it is the majority carriers in the substrate which are captured and re-emitted and the technique therefore measures the properties of majority carrier traps which exist in the substrate. Shown in Figure 4.2 is a typical enhancement n-channel MOSFET with the applied voltages and the resulting drain current transients. During the "off" state when the channel is in accumulation, majority carriers from the substrate rush in and fill any deep traps which may be present. In the ensuing "on" state $(V_G > V_T)$ the drain current which results from the formation of the channel will be affected by the change in the total depletion region charge. It was shown in the previous section that occupied deep states in a depletion region give up trapped charge through the process of trap emission. This trap emission changes the total charge in the depletion region and will therefore modify the drain current. In addition to the switching of the transistor, the temperature of the device is slowly varied during the measurement. This temperature variation results in a number of drain current transient curves which can be monitored and analysed to extract the trap parameters.

The discussion that follows is developed in terms of a donor level in an n-channel transistor with total concentration N_T. These equations will also be applicable to acceptor levels in p-channel transistors with a proper change of notation.

The threshold voltage V_T of a bulk Si enhancement MOSFET neglecting any effect of interface states and fixed oxide charge is

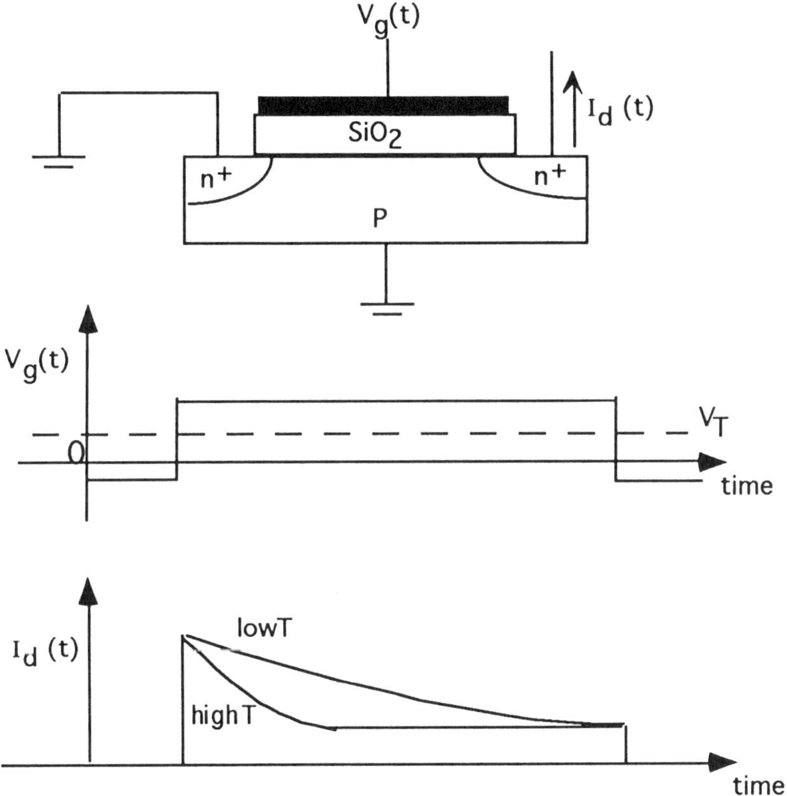

Figure 4.2: Enhancement MOSFET with the applied voltages and derived currents for current DLTS

$$V_T = V_{FB} + 2\Phi_f - \frac{Q_D}{C_{ox}} \tag{4.14}$$

where

$$Q_D = -\sqrt{4\epsilon_{si}q\Phi_f N_A} \tag{4.15}$$

Here V_{FB} is the gate voltage required to establish the flat band condition, ϵ_{si} the permittivity of silicon, N_A the substrate doping (or effective doping in the case of an inhomogenous distribution), and $2\Phi_f$ the surface potential required for strong inversion. The gate oxide capacitance per unit area C_{ox} is defined by $C_{ox} = \epsilon_{ox}/t_{ox}$, where ϵ_{ox} is the permittivity of the gate oxide and t_{ox} the gate oxide thickness. A complete derivation of Equation 4.14 can be found in Reference 17. Figure 4.3 illustrates the energy band diagram of a MOS structure with deep donor states. If $n_T(x)$ represents the total number of unionized (or

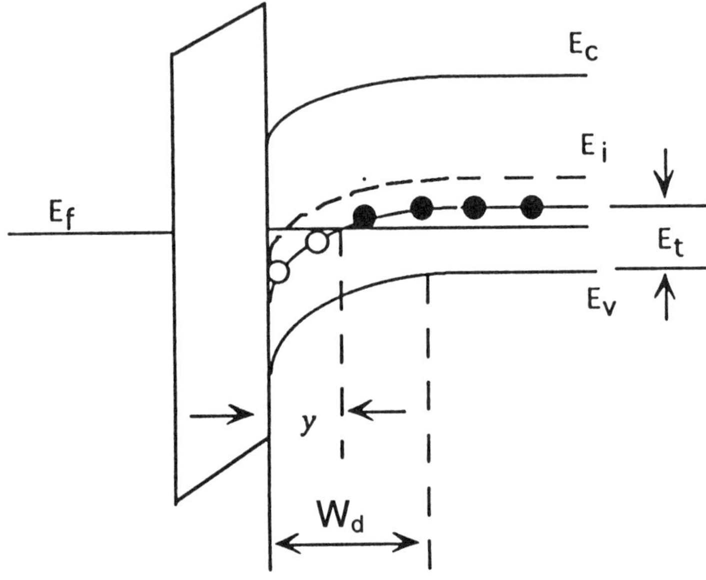

Figure 4.3: Energy band diagram of a MOS structure with a deep donor state at E_t above the valence band edge

occupied) donor states, then the total depletion charge immediately after the transistor is switched to the "on" state can be written as:

$$Q_D = -q \int_0^{W_d(t)} [N_A - n_T(x)]\partial x \qquad (4.16)$$

Using Equation 4.16, the expression for threshold voltage given in Equation 4.14 becomes:

$$V_T = V_{FB} + 2\Phi_f - \frac{q}{C_{ox}} \int_0^{W_d(t)} [N_A - n_T(x)]\partial x \qquad (4.17)$$

As the trapped charge is emitted, the change in the threshold voltage can be found from (4.17) and is given by:

$$\partial V_T \approx -\frac{q}{C_{ox}} [(N_A - N_T)\partial W_d(t) - y\partial n(t)] \qquad (4.18)$$

Here y represents the point where the Fermi level crosses the trap level. In most cases the contribution coming from the shift in the position of the edge of the depletion region can be neglected and the threshold voltage change reduces to [18]:

$$\partial V_T \approx \frac{q}{C_{ox}} y \partial n(t) \tag{4.19}$$

In the above expression, $\partial n(t)$ represents the change in trap concentration from times $t = 0$ to $t = t$. Using Equation 4.12, the change in threshold voltage due to an initially filled trap with concentration N_T can now be written in terms of the trap emission rate as:

$$\partial V_T = \frac{q}{C_{ox}} y N_T \left[1 - \exp(-e_p t)\right] \tag{4.20}$$

For low values of drain voltage the drain current flowing in the transistor in the "on" state can be approximated by [17]:

$$I_D = \mu C_{ox} \frac{W}{L} (V_G - V_T) V_D \tag{4.21}$$

In the current transient experiment the gate and drain voltages are held constant during the "on" state. The observed change in drain current will therefore be entirely due to the change in threshold voltage. Substituting Equation 4.20 into Equation 4.21 results in the following expression for the change in drain current

$$\Delta I_D = -\mu C_{ox} q \frac{W}{L} y N_T \left[1 - \exp(-e_p t)\right] \tag{4.22}$$

Following the approach in Reference 18, the above equation can be written in terms of the transconductance g_m if it is assumed, to first order, that the mobility μ is independent of the gate voltage V_G. The resulting drain current change can now be written as:

$$\Delta I_D = -\frac{g_m}{C_{ox}} q y N_T \left[1 - \exp(-e_p t)\right] \tag{4.23}$$

The anaylsis of Equation 4.23 to extract the trap parameters can be carried out in a number of different ways [6]. For its convience and ease of implementation, the author prefers the double boxcar method which is discussed below.

4.4 Signal Analysis

The output of a double boxcar signal processor is the averaged difference of the input signal taken at two predetermined times t_1 and t_2. This processor can be applied to the drain current transients of the enhancement MOSFET to determine the trap parameters. Illustrated in Figure 4.4 are typical drain current transients obtained at different temperatures. As shown there, at low temperatures the drain current transient approaches the equilibrium value very slowly while at high temperatures the approach is much more rapid. If this signal is fed into a double boxcar averager with the sampling times as indicated in Figure 4.4(a), the output of the boxcar is as shown in Figure 4.4(b). Here the

118

signal is observed to pass through a maximum at a specific temperature T_m. The relationship between the sampling times, peak temperature, and trap emission rate can be derived by considering the conditions under which a maximum output signal is obtained. The output signal of the double boxcar $B(t)$ is given by:

$$B(t) = \Delta I_D(t_2) - \Delta I_D(t_1) = \frac{g_m}{C_{ox}} q y N_T \left(e^{-e_p t_2} - e^{-e_p t_1} \right) \tag{4.24}$$

This signal will pass through a maximum when $\partial B(T)/\partial T = 0$. At this maximum temperature it can easily be shown that the emission rate e_p is related to the sampling times by :

$$e_p = \frac{\ln(t_1/t_2)}{t_1 - t_2} \tag{4.25}$$

This relationship allows one to determine the emission rate corresponding to the peak temperature T_m. If the measurement is repeated with different values of sampling times, a series of emission rate and temperature pairs can be obtained. An examination of Equation 4.9 reveals that by constructing an Arrhenius plot of $\ln(e_p)$ versus $1/T$, the trap energy E_T and capture cross section σ_p can be extracted from the slope and the intercept of this plot respectively. This signal processing method is the boxcar implementation of the rate window concept as outlined by Lang in Reference 7. For a uniformly distributed trap, the concentration can be obtained from Equation 4.24 and is given by:

$$N_T = \frac{C_{ox}}{g_m} \frac{B(T)}{q y \left(e^{-e_p t_2} - e^{-e_p t_1} \right)} \tag{4.26}$$

Here $B(T)$ is the height of the current transient peak and g_m the transconductance measured at the peak temperature and "on" state gate voltage. Using the usual parabolic variation of the potential in the depletion region, y can be expressed as:

$$y = W_d - \sqrt{\frac{2\epsilon_{si}}{q^2}(N_A - N_T)(E_T - E_F)_{bulk}} \tag{4.27}$$

where W_d is the depletion region width, and $(E_T - E_F)_{bulk}$ is the energy difference between the level of the trap and the Fermi level in the quasineutral region of the semiconductor. For a nonuniform trap distribution, a concentration profile can be obtained from [18]:

$$N_T(y) = \frac{C_{ox}}{q \left(e^{-e_p t_2} - e^{-e_p t_1} \right)} \frac{\partial}{\partial y} \left(\frac{B(T)}{g_m} \right) \tag{4.28}$$

In this measurement, y is varied by changing the depletion width W_d using the substrate bias.

Illustrated in Figure 4.5 and Figure 4.6 are current DLTS spectra obtained from a bulk MOS transistor and a SOI transistor respectively. The bulk n-channel MOSFET had a gate oxide thickness of 120 nm and a substrate doping

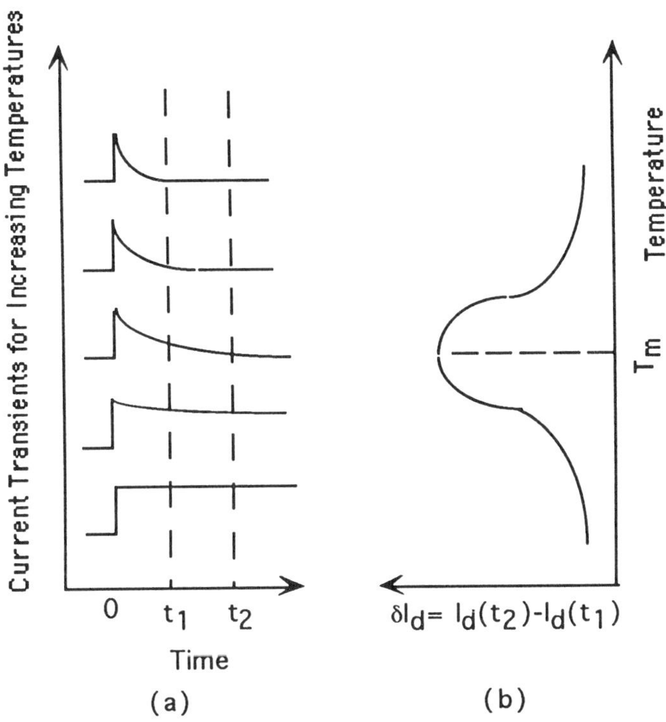

Figure 4.4: The rate window concept using a double boxcar integrator. The output (b) is the average difference of the current transients sampled at times t_1 and t_2 at various temperatures (after Lang, Ref. 7)

of 1.5×10^{15} cm^{-3}. The current transient spectrum was recorded between 100 and 300 K with "off" state and "on" state gate voltages of -3 V and 1 V respectively. During the measurement the drain voltage was held constant at 50 mV. Analysis of a number of such spectra obtained with different double boxcar sampling times indicates the presence of a hole trap at 0.3 eV above the valence band edge with a capture cross section σ_p of 10^{-14} cm^2. From the height of the spectra, the trap concentration was calculated using Equation 4.26 to be 6×10^{13} cm^{-3}. For calculation of the trap concentration the transconductance must be measured at the peak temperature. The SOI transistor was fabricated on a SIMOX substrate with a silicon substare thickness of about 100 nm. The devices were fully depleted in the "on" state and had dimensions of gate widths and lengths of 50 μm and 10 μm respectively. The gate oxide thickness was 25 nm, and the substrate doping was 3×10^{16} cm^{-3}. During the current DLTS measurement on these devices special care was taken to eliminate the effect of the back SiO_2/Si interface. As shown in Figure 4.6, the back interface could be effectively screened during the measurement by applying a substrate voltage

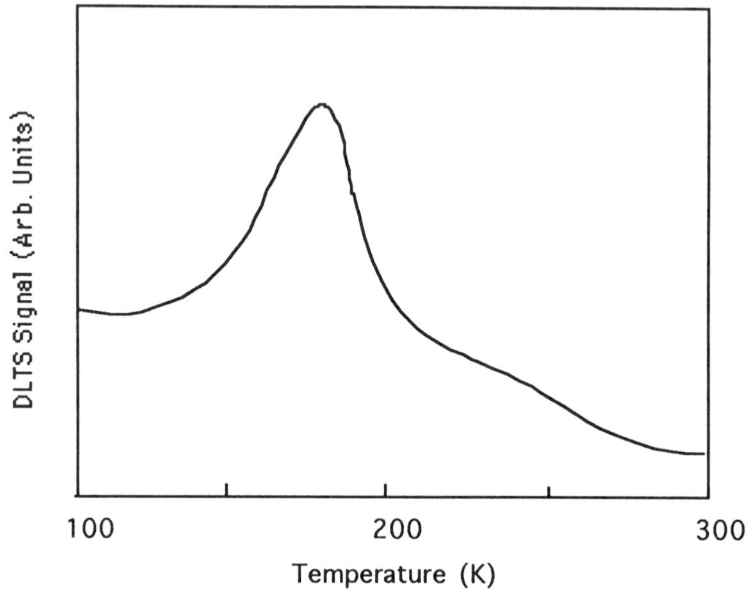

Figure 4.5: MOSFET current DLTS spectrum for a n-channel transistor with drain bias $V_D = 50\ mV$ and gate pulses between -3 V and 1 V. (after Bauza et. al., Ref. 18)

$V_b = -10\ V$ which held this surface in accumulation. With this accumulation voltage present, the front gate was switched between 0 V and 1.3 V with a constant drain voltage of 50 mV. Analysis of the spectra resulted in a hole trap at 0.44 eV above the valence band edge with a capture cross section of $10^{-15}\ cm^2$.

Further refinements to the above model have been carried out in Reference 18. There the full resolution of the Poisson equation assuming the presence of a deep level impurity has been considered. The variation of the depletion layer width, which was neglected above is also included. The major improvement over the simpler model is in the determination of trap concentration. It has been shown that the simpler model can underestimate the trap concentration by as much as 30%.

Before leaving the subject of enhancement current transient spectroscopy there is a variation of the above measurement that should be mentioned. In this technique the transistor is pulsed from strong inversion to weak inversion while monitoring the voltage required to keep the drain current constant (the drain current could also have been measured without changing the result). The drain current is observed to vary in steps as compared to the exponential transients obtained above. In the analysis presented in Reference 19, the steps are

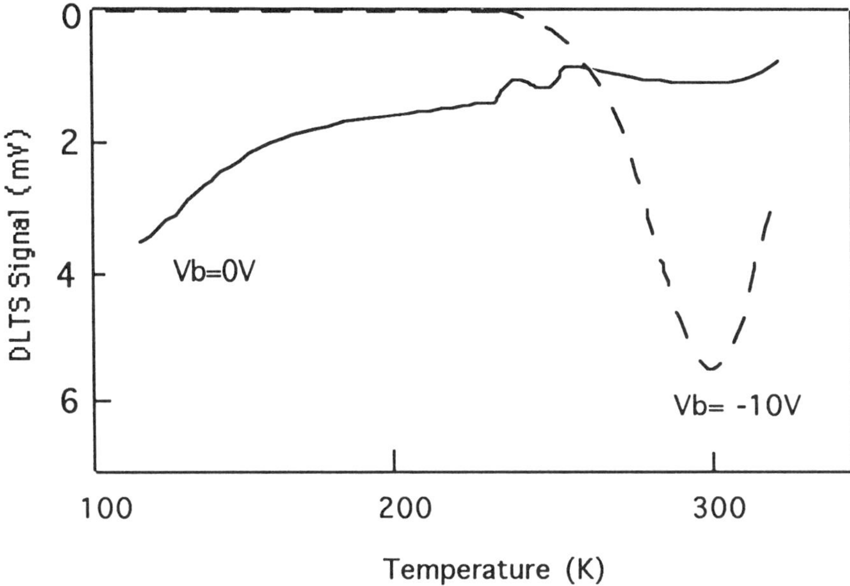

Figure 4.6: MOSFET current DLTS spectrum for a fully depleted SOI n-channel transistor with drain bias $V_D = 50\ mV$ and gate pulses between 0 V and 1.3 V. (after McLarty et. al., Ref. 16)

averaged and an Arrhenius plot is generated. This technique appears to measure traps located in the oxide and not SiO_2/Si interface and/or silicon bulk traps. The reader is refered to Reference 19 for a more detailed discussion of the technique.

4.5 Depletion MOSFET Current Transient Spectroscopy

In depletion MOSFET current transient spectroscopy the transistor is usually pulsed from accumulation to a gate voltage that is sufficient to form a depletion region. A typical depletion mode MOSFET structure is shown in Figure 4.7. The transistor is usually isolated by a pn junction or an insulating layer (eg. SOI). The drain current then flows between the depletion region edge and the underlying junction or insulating layer. It can be shown that the drain curent of a depletion MOSFET on an insulating substrate is described as [17]:

$$I_D(t) = \frac{W(x_t - W_d)}{\rho L} V_D \tag{4.29}$$

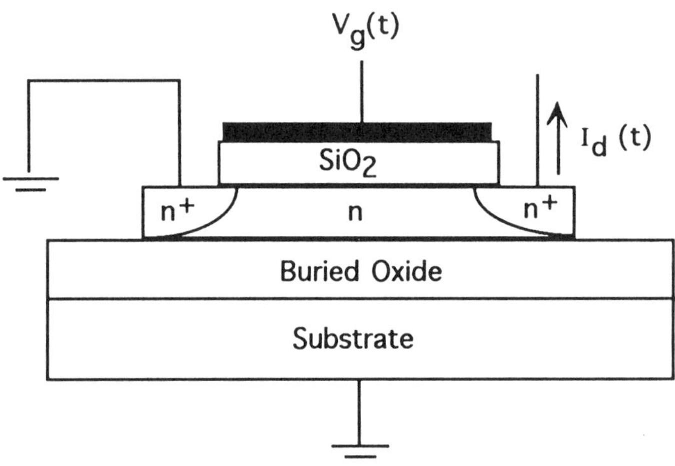

Figure 4.7: Delpetion mode SOI transistor

where ρ is the channel resistivity (assumed constant), W and L the channel width and length respectively, x_t the film thickness, and W_d the depletion layer width.

If traps are present in the depletion region they will be filled with majority carriers when the transistor is pulsed into accumulation. In the ensuing depletion region the traps will give up there charge and change the width of the region. It can be shown that the rate of change of the depletion region width is given by:

$$W_d(t) = \frac{2\epsilon_{si}\Psi_s}{qN_A}\left[1 + \frac{N_T}{2N_A}\exp\left(-\frac{t}{\tau}\right)\right] \tag{4.30}$$

Substituting Equation 4.30 into Equation 4.29 results in an expression that describes the drain current transient in terms of the emission rate. If this drain current is used as the input signal to a double boxcar signal processor the output signal will pass through a peak at a particular temperature T. As in the case of the enhancement MOSFET analysis discussed earlier, the emission rate at the peak temperature is determined from the setting of the boxcar processor. If the observed peak is due to the presence of discrete traps in the bulk, constructing an Arrhenius plot allows the extraction of the properties of the deep state. The presence however of the Si/SiO_2 interface in the depletion region results in a complicated signal that has contributions from a continum of states at this interface as well as discrete bulk traps. In some cases it is possible to distinguish between the states as the signal from the interface states is usually a broad slowly varying featureless spectrum, while the spectra due to bulk traps are sharply defined. In cases where it is possible to differentiate the signals the

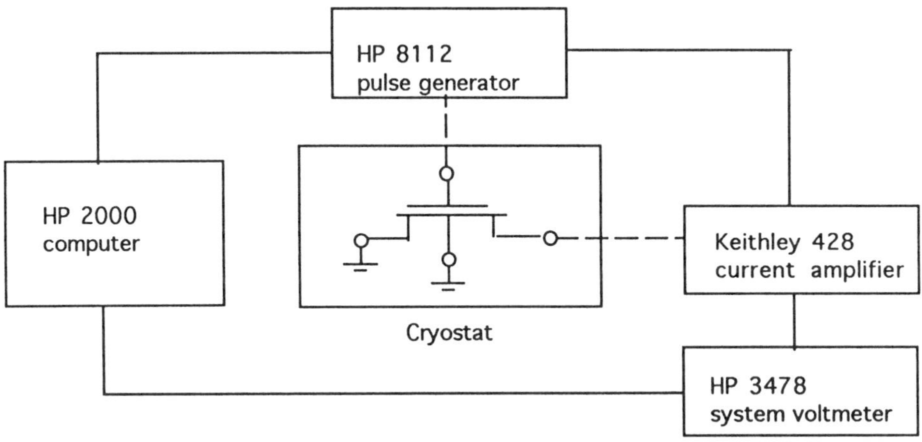

Figure 4.8: MOSFET current DLTS system

concentration of the bulk traps can be found from Equations 4.29 and 4.30. In these measurements it is important to keep the gate voltage below that required to form an inversion layer. If the gate voltage during the measurement phase exceeds this voltage an additional component due to minority carrier generation will be introduced into the measured signal. If there are no observable deep states present in the substrate, this generation signal can be analysed using the rate window concept to provide information on the generation rate at different temperatures as well as the dominant generation mechanism [20].

4.6   MOSFET Current DLTS Measurement System

The MOSFET current DLTS measurement system is very similar to the standard capacitance DLTS system described in the literature [6]. Here the major difference is the substitution of a sensitive current meter for the capacitance meter. Shown in Figure 4.8 is a system that can be used to perform these measurements. As illustrated in Figure 4.8, the system consists of four major components: a cryostat with a temperature range of 80 K to 450 K with an accuracy of ± 0.1 K, a Keithley 428 current amplifier, a HP 3478 voltmeter, a HP 8112 pulse generator and a HP 2000 computer. Each component is controlled by the computer which also sets the timing of the measurement. The sequence of the measurement is as follows: the temperature of the sample is first set by the computer and held constant while a series of drain current tran-

124

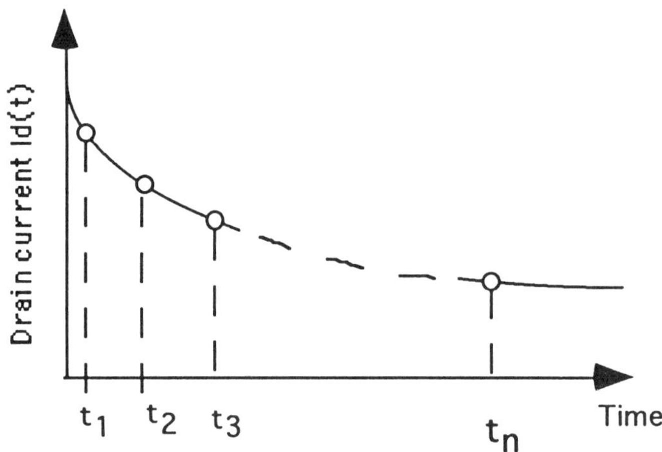

Figure 4.9: Drain current transient showing sampling times

sients are recorded at this temperature. With the temperature held constant, the gate voltage is pulsed from inversion to accumulation and back to inversion. While in inversion the entire resulting drain current transient is measured by the Keithley curent amplifier. These transients are averaged by the computer and stored for later signal analysis. The boxcar processor discussed earlier is implemented using the data points in the computer. An important advantage of storing the entire transient is that only one temperature scan is required as multiple rate windows can be used at each temperature point.

This is illustrated in Figure 4.9 where the typical sampling points for a drain current transient are shown. Each pair of points can be used to represent a rate window by forming the difference signal. For example, at the points t_1 and t_2 there are currents measured over the entire temperature range. The difference signal $I_D(t_2) - I_D(t_1)$ would then be plotted as a function of temperature to obtain a spectrum and represent one rate window. By repeating this process for other pairs of points a series of spectra and the resulting Arrhenius plot can be obtained from a single temperature scan.

Using the above system, a typical experiment takes about 50 minutes with each curve averaged 100 times. The temperature step used is typically 1 K and the system has a sensitivity of $N_T = 10^{-5} N_A$.

Bibliography

[1] W. Kohn and J.M. Luttinger, *Phys. Rev.*, 98, pp. 883-888 (1955).

[2] S. T. Pantelides, *Solid State Comm.*, 14, pp. 1255-1258 (1974).

[3] D.V. Lang, *Topics in Applied Physics: Thermally Stimulated Relaxation in Solids*, P. Braunlich, ed., 37, Springer, Berlin, pp. 93-133 (1979).

[4] D.K. Schroder, *Semiconductor Material and Device Characterization*, Wiley-Interscience, John Wiley and sons, New York, pp. 297-358 (1990).

[5] C.T. Sah, *Solid State Electronics*, 19, 975 (1976).

[6] G.L. Miller, D.V. Lang, and L.C. Kimerling, *Ann. Rev. Mater. Sci.*, pp. 377-448 (1977).

[7] D.V. Lang, *J. Appl. Phys.*, 45, no.7, pp. 3023-3032.

[8] B.W. Wessels, *J. Appl. Phys.*, 47, pp. 1131-1133 (1976).

[9] J.A. Borsuk and R.M. Swanson, *IEEE Trans. Electron Dev.*, ED-27, pp. 2217-2225 (1980).

[10] Y. Tokuda and A. Usami, *Japan J. Appl. Phys.*, 22, 371 (1983).

[11] I.D. Hawkins and A.R. Peaker, *Appl. Phys. Lett.*, 48, pp. 227-229 (1986).

[12] P.K. McLarty, D.E. Ioannou, and H.L. Hughes, *Appl. Phys. Lett.*, 53, pp. 871-873 (1988).

[13] D.P. Vu, A. Chantre, D. Ronzani, and J.C. Pfister, *Proc. Mat. Res. Soc. Symp.*, 53, pp. 357-361 (1986).

[14] M.G. Alderstein, *Electron. Lett.*, 12, pp. 297-298 (1976).

[15] J.W. Chen, R.J. Ko, D.W. Brzezinski, L. Forbes, and C.J. Dell'Oca, *IEEE Trans. on Elect. Dev.*, ED-28, pp. 299-304 (1981).

[16] P.K. McLarty, D.E. Ioannou, and J.P. Colinge, *IEEE Electron Dev. Lett.*, EDL-9, pp. 545-547 (1988).

[17] S.M. Sze, *Physics of Semiconductor Devices*, Wiley-Interscience, John Wiley and Sons, New York (1981).

[18] D. Bauza and G. Ghibaudo, *J. Appl. Phys.*, 70, pp. 3333-3337 (1991).

[19] A. Karwath and M. Schulz, *Appl. Phys. Lett.*, 52, pp. 634-636 (1988).

[20] P.K. McLarty and D.E. Ioannou, *IEEE Trans. on Elect. Dev.*, ED-37, pp. 262-266 (1990).

5

Individual Interface Traps and Telegraph Noise

H. H. Mueller and M. Schulz
University of Erlangen
Erlangen, Germany

5.1   Introduction

The immense progress in silicon technology and miniaturization of electronic devices to sub-micrometer sizes has led to almost defect-free metal oxide semiconductor field effect transistors (MOSFETs). The centerpiece of microelectronics technology, the $Si - SiO_2$ interface, is today fabricated with an interface trap density $D_{it} = 10^8 - 10^{10}$ cm^{-2}eV^{-1}, so that a standard MOSFET with sub-μm gate dimensions contains less than 1-100 defects in its active area. A schematic of an n-channel MOSFET is depicted in Figure 5.1(a). The carrier densities induced in the channel by the variable gate bias voltage range from $n_s = 10^9$ cm^{-2} in the sub-threshold bias region up to $n_s = 10^{12}$ cm^{-2} for high gate bias voltages. The number of charge carriers in the channel of a micrometer-sized MOSFET therefore ranges from about 10 in the sub-threshold region up to more than 10000 electrons in strong inversion. Capture of a single electron into an interface trap causes a noticeable 0.01% to 10% change in the number of mobile charge carriers in the channel and thus in its conductance. Conductance changes of the channel may be even larger, because the mobility is also affected

128

by the trapping or emission of a charge carrier due to the creation or annihilation of a scattering center. Trapping and re-emission of single electrons from and to the channel by interface traps cause a random switching of the source-drain conductance between two discrete states, as schematically sketched out in Figure 5.1(b). Because of its similarity to a telegraph signal this switching is called random telegraph signal (RTS) or random telegraph noise. Especially for weakly inverted channels near threshold gate bias, the conductance change induced by single-electron trapping in the active gate area is easily observed at room temperature and may be used to study and characterize individual defects at the $Si - SiO_2$ interface.

Not only the number of interface traps present is scaled with the magnitude of the device size ; the potential perturbation induced in the interface due to the traps is also affected. This effect is simply demonstrated by estimating the distance r_C at which the Coulomb potential of a point charge decays into the thermal background, *i.e.* $V \simeq 0.5kT/q \simeq 12\,mV$ at room temperature. This distance is about $r_C \approx 30\,nm$ or more for lower temperatures. Note that the mean distance between traps is $d \approx 58\,nm \approx 2rC$ for a spatial interface trap density $N_{it} = 3 \times 10^{10}\,cm^{-2}$. The trap potential perturbation therefore fills the interface area without gap and may be represented by a mean potential, e.g. a uniform flat-band voltage shift. In small MOSFETs having thin oxides ($t_{ox} \leq 10\,nm$) the gate electrode, having metal-like conductivity, is very close compared to r_C to the trap centers and thus screens the trap potential. The typical screening length r_{scr} in this case is the oxide thickness $r_{scr} \approx t_{ox}$. The extent of the trap potentials is reduced by a factor of 3 to the order of the oxide thickness. The impact of scaling the oxide thickness is depicted in Figure 5.2. For the classical large MOSFET, the oxide thickness is of the order of $t_{ox} \approx 30\,nm$ or more. In this case (see Figure 5.2(a)), the trap potentials overlap and may be represented by the average. However, for typical sub-μm MOSFETs the oxide thickness is reduced to $t_{ox} \approx 10\,nm$. This case is depicted in Figure 5.2(b), where the traps are fully isolated and de-coupled.

The down-scaling of MOSFETs to sub-μm sizes therefore has two effects: the number of defects is cut down to a few single traps and the traps become isolated and de-coupled from each other. Single, individual traps and their properties may therefore be studied and characterized in sub-μm MOSFETs.

RTSs have been found in many devices other than MOSFETs, such as in MOS tunnel diodes [1], in tunneling microscopes [2], in MESFETs [3] etc. Only for the MOSFET, a broad literature exists and a modeling of the RTS trapping effects with detailed balance of the rates is possible, because of the perfect insulation by the gate oxide. The RTS is responsible for the high noise levels in micron-sized MOSFETs [4]. It has also been demonstrated that the superposition of many RTSs is the cause of the $1/f$ power spectrum. The RTS permits to study for the first time the capture and emission kinetics of MOS interface states on single, individual defects, in contrast to the capacitance-voltage or conductance techniques, where a mean value of the time constants of a large capacitor is measured and the time constants are distributed by surface potential fluctu-

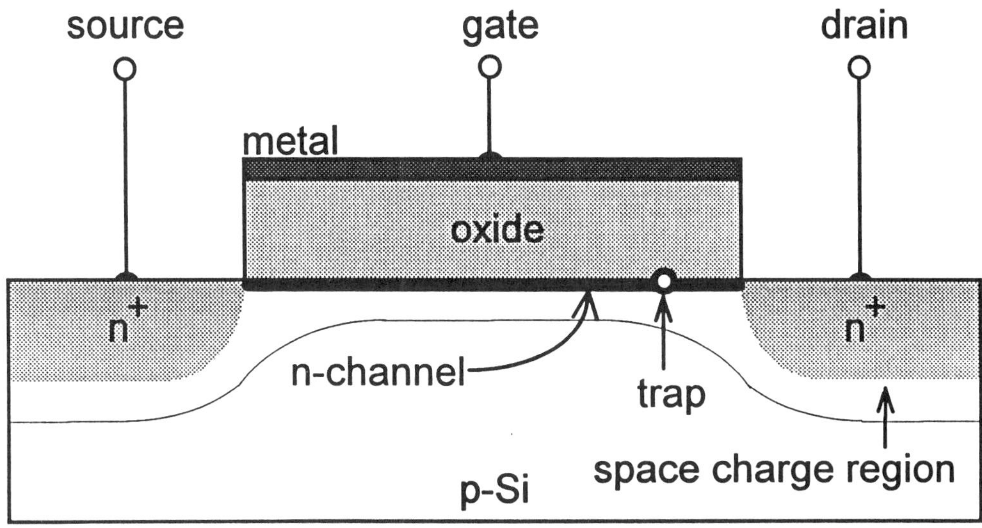

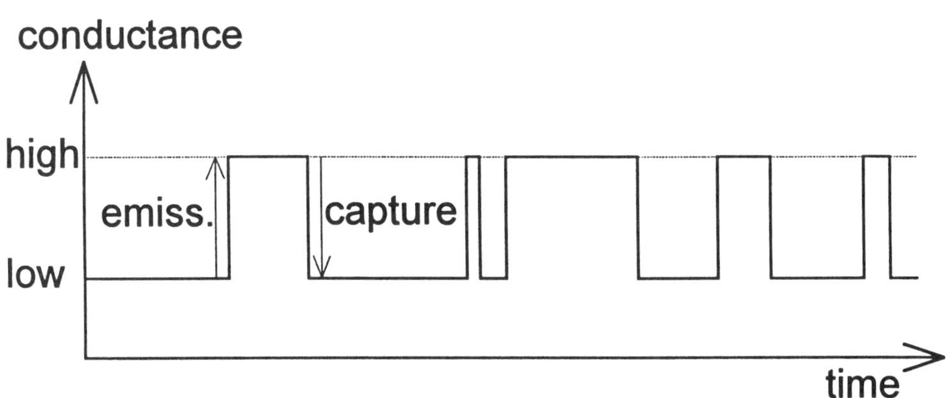

Figure 5.1: (a) Schematic of an MOS transistor (n-channel device). A single interface trap is indicated in the active gate area. (b) Schematic of a random telegraph signal. The source-drain conductance is shown as a function of time. Emission and capture processes are indicated by arrows from the low to the high state and vice versa, respectively.

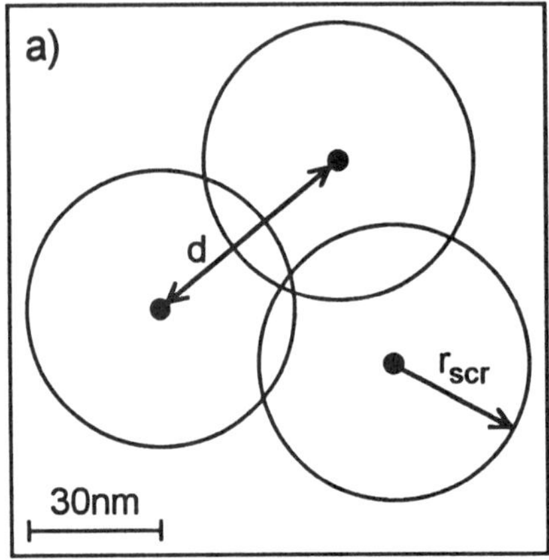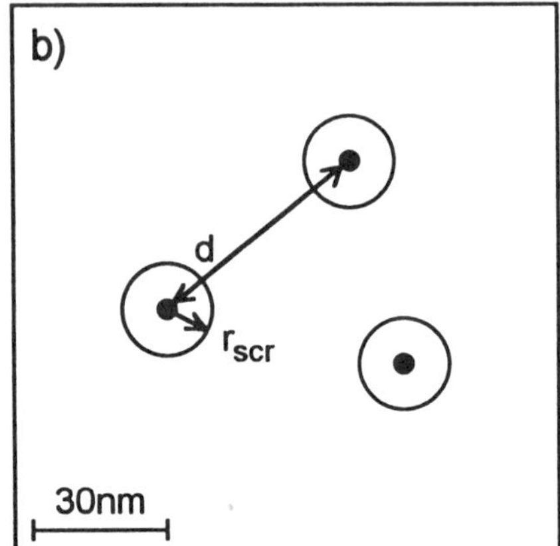

Figure 5.2: Extent of potential perturbations around randomly placed point charges in the $Si - SiO_2$ interface at mean distance $d \approx 58$ nm, *i.e.* $N_{it} \approx 3 \times 10^{10}$ cm^{-2}. The arrows r_{scr} indicate the screening radius $r_{scr} \approx t_{ox}$. a) The potentials of point charges overlap for thick oxides $t_{ox} > d/2$. The figure depicts screening radii for $t_{ox} \approx 30$ nm, as commonly used in classical large transistors. b) The potentials of point charges are isolated for thin oxides $t_{ox} < d/2$. The figure depicts screening radii for $t_{ox} \approx 10$ nm as frequently used in sub μm MOSFETs.

ations. The characteristic time constants are determined for one specific defect level. New results and unexpected properties of the traps at the MOS interface have been revealed by these measurements. Our present understanding of the RTS properties generally observed and the models describing these properties are presented in the following sections together with a comprehensive literature review.

5.2 Observation of Single Carrier Trapping

5.2.1 Random Telegraph Signals

When monitoring the source-drain conductance of the MOSFET channel, an RTS like that schematically shown in Figure 5.3 may be observed. The four oscilloscope traces in Figure 5.3 of the RTSs are taken at different gate voltages increasing from bottom to top. The main effect of the gate voltage is to increase the carrier density in the channel, which is monitored by the drain current I_D induced for a drain bias of $V_D = 10$ mV. This bias voltage is kept small, less than the thermal voltage kT/q, in order to avoid a gradient in the channel

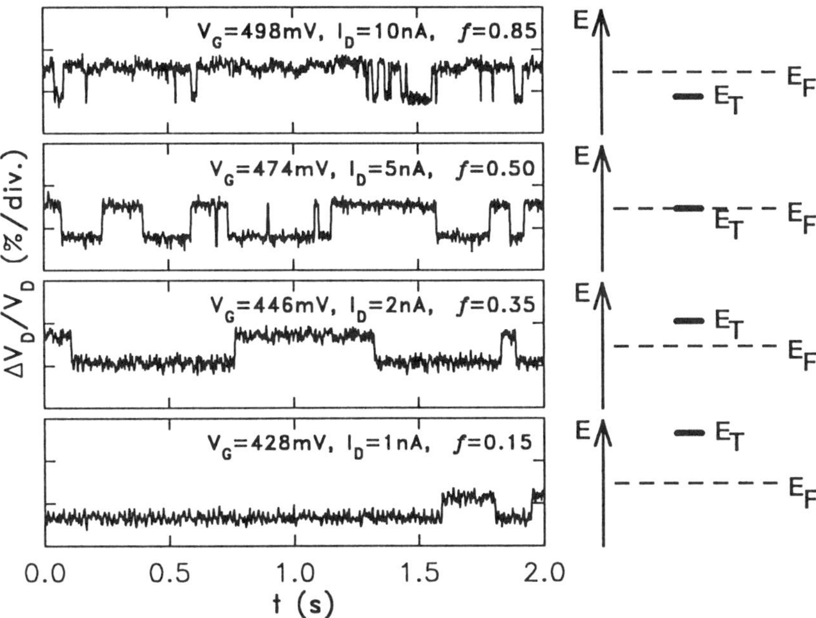

Figure 5.3: Random telegraph signals observed in the source-drain resistance of a sub-μm MOSFET, depicted as a relative voltage change $\Delta V_D/V_D$ at $V_D = 10$ mV for constant drain current I_D and gate voltage as indicated. The switching is due to a single interface trap level. The Fermi occupation factor f listed in the inserts varies according to the relative position of the trap level to the Fermi energy as depicted on the right.

carrier density induced by the lateral field. It is noted that the magnitude of the RTS conductance switching step observed typically ranges around 1%, as estimated by the number change above; however, step heights of up to 30% have been observed.

The discrete switching is attributed to the capture and emission of a single inversion charge carrier in the channel into and out of a trap level. The trap occupation can be extracted from the RTS by the mean time durations the charge carrier spends in (τ_e, emission) and out of (τ_c, capture) the trap level. The Fermi occupation factor is

$$f = \frac{\tau_e}{\tau_e + \tau_c} = \frac{1}{\exp\left(\frac{E_T - E_F}{kT}\right) + 1} \tag{5.1}$$

The trap energy position E_T with respect to the Fermi energy can be determined. The occupation factors f and the corresponding relative positions of the trap energy and Fermi energy are indicated for the four RTS traces in

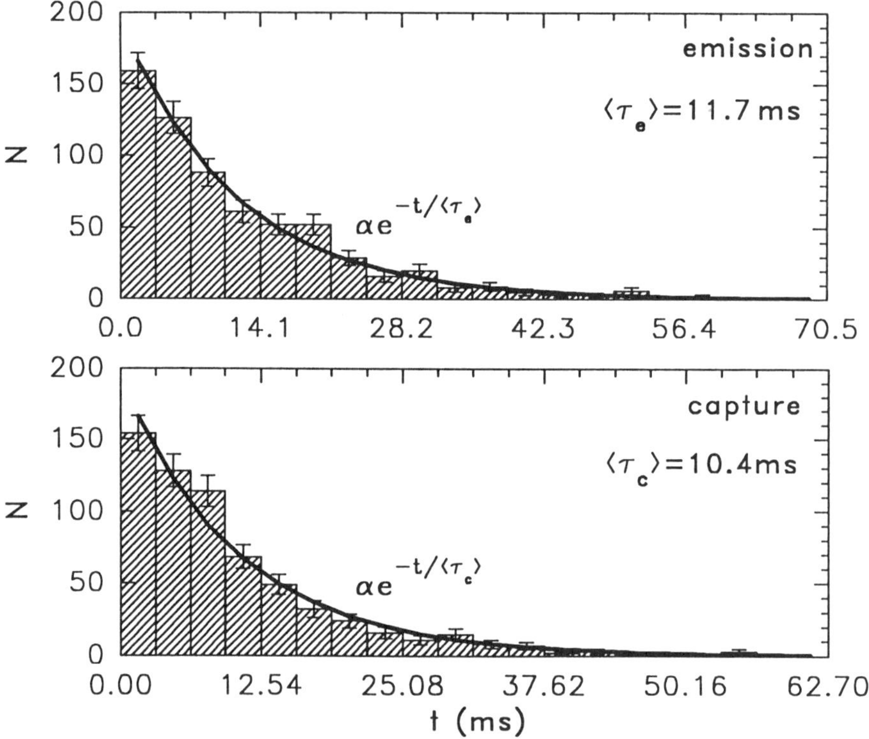

Figure 5.4: Typical histograms of capture and emission time distributions measured for a specific trap. The curves indicated are exponential dependencies, in good agreement with Poisson distributions. The top graph is for emission, the bottom graph for capture.

Figure 5.3. For an increased carrier density in the channel the trap causing the RTS in Figure 5.3 is lowered from a position above the Fermi energy to a position below it.

Unlike the situation of an ensemble of traps, where a fractional occupation results in some traps being full and some traps being empty at the same time, a fractional trap occupation for a single trap considered in an RTS is determined by a time average. In accordance with the ergodic theorem, which equates time and ensemble averages, the same occupation factor is determined.

Discrete fluctuations in the source-drain conductance of small MOSFETs similar to those in Figure 5.3 were first observed and investigated by Ralls and co-workers in 1984 [5] who looked into the 1/f noise of such devices. The noise power has been shown to be proportional to the gate area and thus to the number of RTSs present. RTSs are resolved in small devices over several orders in drain current from depletion to moderate or strong inversion. The chemical nature of the traps involved is not yet clear. Quite a variety of different microscopic phenomena, such as dangling bonds and attractive, neutral or repulsive impurities may lead to similar macroscopic appearances of an RTS.

During the last decade this phenomenon has been studied extensively for the $Si - SiO_2$ interface from liquid helium to room temperature [5, 6, 7], using n- and p-channel MOSFETs. Single-carrier trapping and the phenomena associated with it have been previously reviewed by Kirton and Uren [4], Schulz and Karmann [8], and Farmer [9].

5.2.2 Emission and Capture Times

RTSs provide a unique opportunity to measure the capture and emission times of single charge carrier traps. This is done by averaging over the time intervals a signal like the one in Figure 5.1(b) spends in the high or low conductance state. A problem is the attribution of the switching direction to either emission or capture. For some RTSs the switching direction is opposite to that shown in Figure 5.1(b). The time constants are correctly attributed to emission and capture by their gate bias dependence in such a way that the mean trap occupation f defined by Equation 5.1 increases with increasing carrier concentration in the channel. Typical dependencies of the mean capture and emission time constants will be discussed for the different types of traps in a section further below. The switching times are found to be exponentially distributed (Figure 5.4). Both emission and capture of a charge carrier are therefore random processes without memory, *i.e.* the probability of emission from a filled trap or capture into an empty trap within the next interval of unit time is constant, regardless of history [10]. The exponential decay times represent the mean emission or capture time constants, respectively. The standard deviation, which for an exponential distribution is equal to the mean value, may be used to test the accuracy of a measurement.

5.2.3 Emission Transients

Another method by which one can directly observe the carrier emission from traps in small MOSFETs and identify the conductance change of emission processes is deep level transient spectroscopy (DLTS) [11, 7]. In this measurement method, all the traps are filled by biasing the MOSFET into strong inversion, re-emission transients are then observed after switching the gate bias voltage back into weak inversion. Typical examples of such transients (all sequentially taken from the same sample) are presented in Figure 5.5.

The emission transients are quantized due to the low number of traps involved. The number of traps may be determined by counting the number of steps. Only 4-6 traps are visible in Figure 5.5. Due to the random time durations for emissions, the step distribution is different for each transient. It is noted, however, that the step height varies from trap to trap due to the different modulation of the drain current by the various traps. Individual traps can

134

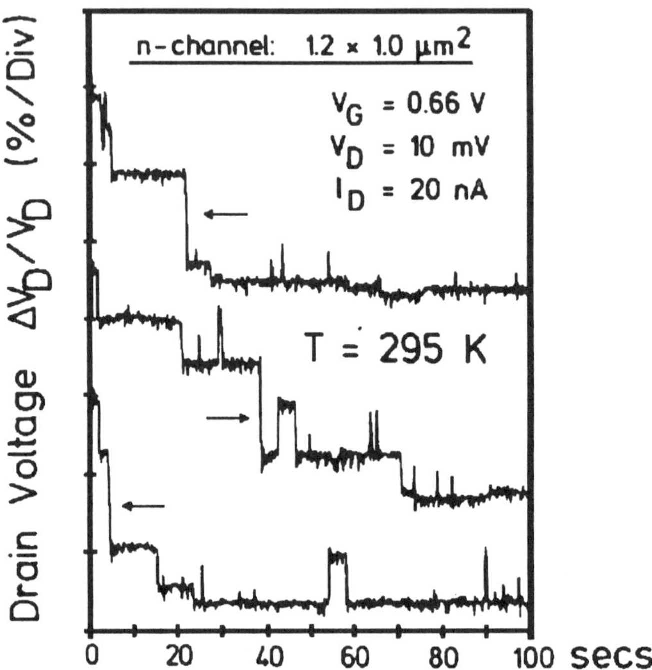

Figure 5.5: Quantized DLTS-type emission transients for an n-channel MOS-FET after complete filling of the interface traps. The plot shows the relative voltage drop across the channel vs. time for three sequential transients (vertically displaced to avoid superposition) taken at the same conditions. The arrows mark the occurence of the same step height.

therefore be identified in the different transients by their step height although in some cases the order of the emission events changes. For most traps (about 95%), the channel conductance increases at elevated temperatures above 77 K after an emission event; for a few traps, a decrease in the channel conductance is observed; in some cases, a correlated emission of several charge carriers takes place. These complex RTSs will be discussed later. The DLTS principle also provides a good alternative method to attach the high and low channel conductance states to the trap occupation states, full and empty.

5.2.4 Noise

The random switching of the source-drain current by RTSs creates noise. The noise power spectrum of an RTS is well understood and frequently discussed in textbooks [12]. The noise power spectrum of a single RTS with one voltage amplitude ΔV is given by the Lorentzian

$$S(\omega) = \frac{4(\Delta V)^2 f(1-f)\bar{\tau}}{1+(\omega\bar{\tau})^2}, \qquad (5.2)$$

with f being the Fermi factor and the effective time constant $\bar{\tau}$ given by

$$\frac{1}{\bar{\tau}} = \frac{1}{\bar{\tau}_e} + \frac{1}{\bar{\tau}_c} \qquad (5.3)$$

It is noted that for the measurement of the noise power spectrum only two parameters, the plateau height $4(\Delta V)^2 f(1-f)\bar{\tau}$ and the cut-off frequency $\omega = \omega_{c0} = 1/\bar{\tau}$ can be determined. A direct measurement of the RTS in the time domain yields three parameters $\bar{\tau}_e$, $\bar{\tau}_c$ and ΔV, so that more information on the trap occupation is available.

A noise power spectrum measured in a sub-μm MOSFET under the bias conditions given is depicted in Figure 5.6.

The noise power spectrum observed with only 3 traps active already approaches a 1/f dependence as indicated by the straight line. The dotted curves I-III, which represent the Lorentzians calculated from RTS measurements of the individual traps, superimpose to the noise power spectrum. It is concluded that the 1/f noise spectrum in large area MOSFETs is generated by a superposition of many Lorentzians differing in the step height and exponentially distributed with respect to the time constant [4, 13].

5.2.5 Complex Random Telegraph Signals

The basic shape of an RTS due to switching of an individual interface trap is depicted in Figure 5.3. Quite commonly the transients will look somewhat more complicated than the simple example presented there. On a closer analysis, however, many of them simply contain the superimposed, independent switching of several traps. The transients are therefore a mere sum of several independent, simple RTSs. Figure 5.7 shows an example for two traps causing independent switching with two different amplitudes.

A certain percentage (about 4%, [10]) of RTSs exhibit coupled switching, where the switching with one amplitude depends on the state of another trap. Coupled switching may occur with a second signal as an envelope (Figure 5.8).

The coupling of RTSs is made plausible by interpreting it as an electrostatic interaction between traps that are located sufficiently close to each other or by sequential capture of two electrons into the same trap [14]. Complex switching may also arise by the competition between two different time constants involved in the charge transfer process for attractive traps. These will be discussed further below. Other, comparatively rare types of complex RTS signals were discussed in detail by Kirton and Uren in Reference [4].

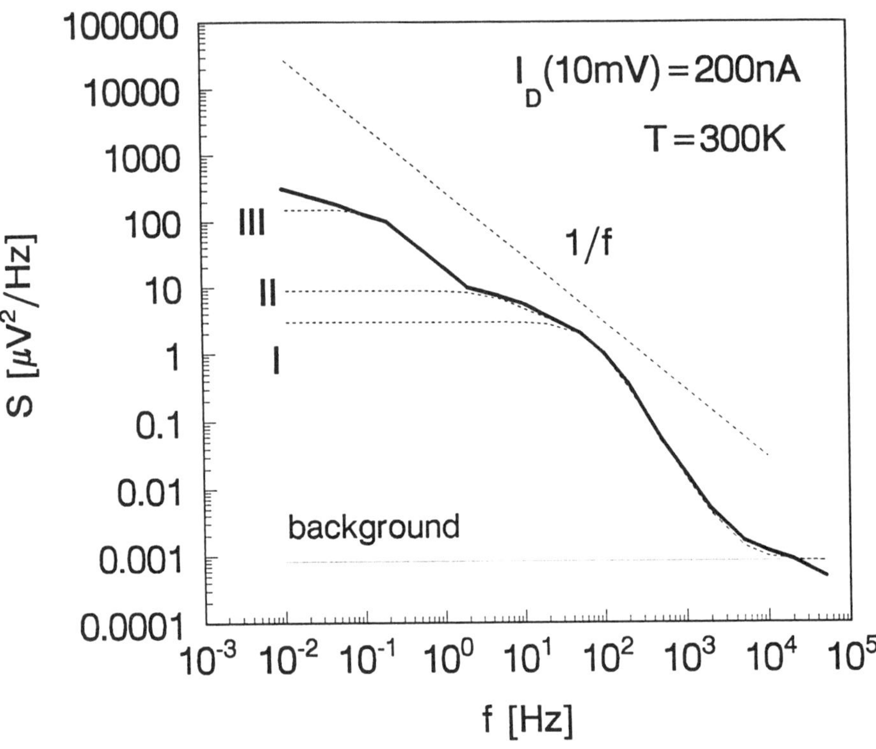

Figure 5.6: Noise power S spectrum of a small area (0.6μm $\times$ 0.7μm) MOSFET showing only three trap levels in the RTS. The full bold curve is the measured spectral noise power. The dashed curves show the Lorentzian power spectra of the three individual traps I-III. The dotted line represents the background noise level. The slope of 1/f noise is also indicated. The insert gives the measurement conditions.

5.3 Experimental Properties of Individual Interface Traps

RTSs in MOSFETs analyze the trapping under equilibrium conditions of a single, individual interface trap. This is important because the trap levels at the interface between the amorphous oxide and the silicon are randomly distributed with respect to energy and time constant. Measurements on a single trap yield well defined parameters which are not averages over an ensemble of many differing traps. The equilibrium conditions permit to relate the emission and capture times to a mean thermal occupation by a detailed balance law similar to Shockley-Read-Hall (SRH) statistics [15]. Three parameters may be measured from an RTS, the mean emission rate, the mean capture rate, and the switching step height, *i.e.* the magnitude of the conductance modulation.

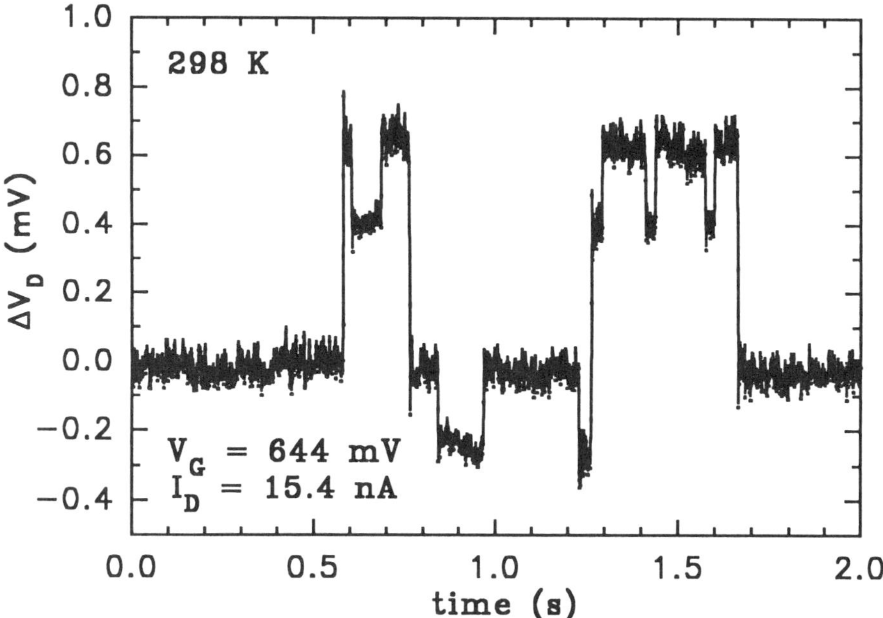

Figure 5.7: Typical RTS that exhibits independent switching with two different amplitudes superimposed. The MOSFET driving conditions are shown in the insert ($V_D = 10$ mV).

These three parameters are dependent on various measurement variables, e.g. the bias conditions (gate, drain, substrate), the temperature and indirectly the carrier density in the channel. The trapping time constants may be related to the carrier density in the channel and to energy level positions similar to SRH statistics. The time constants are therefore studied to a great extent of detail and are well understood. The characteristic dependencies of the time constants may be used to classify the trap types and to model the processes involved. The channel conductance modulation is more complex, because it involves scattering by partially screened trap potentials and because its magnitude is also affected by inhomogeneous current distributions in the finite geometry of the channel. The modulation of the channel conductance is not well understood at present and will only be briefly treated in the following sections.

5.3.1 Classification of Trap Types

The main traps observed with two-level RTSs may be classified into three groups. Basis for the classification is the typical dependence of the mean emission and capture time constants on carrier density in the channel. Examples

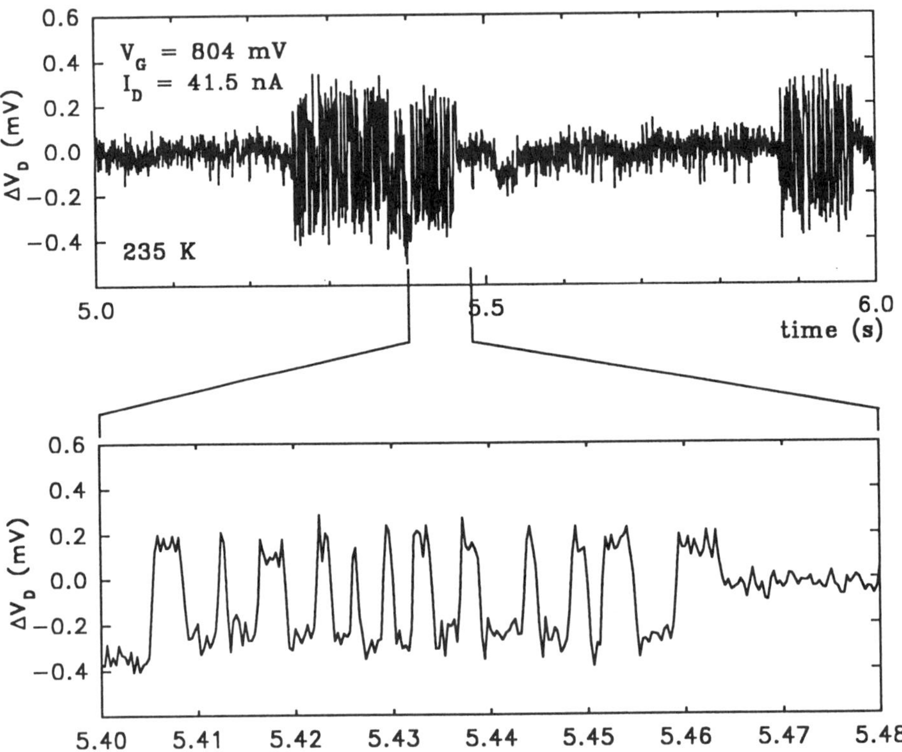

Figure 5.8: Complex RTS for which the rapid switching is interrupted. The expanded interval at the bottom shows the RTS nature of the rapid switching.

of the three classification types are shown in Figures 5.9–5.11. The attribution of the trap types to neutral and attractive traps and a trap bistability is obtained by the theoretical modeling, which consistently describes the observed properties. The model interpretation will be discussed later, for clarity the nomenclature is used here, too. The three types are:

Neutral Trap

The charge state of this trap is neutral when unoccupied by the charge carrier. The trap becomes repulsively charged when a mobile charge carrier from the channel is trapped. The typical dependence of the emission and capture time constants is depicted in Figure 5.9.

The mean emission time is constant or only weakly increasing at high carrier densities ($n_s > 10^{11}$ cm^{-2}) in the channel. The mean capture time decreases steeper than $1/n_s$ with carrier density. This type of RTS is most frequently observed in n-channel as well as in p-channel MOSFETs.

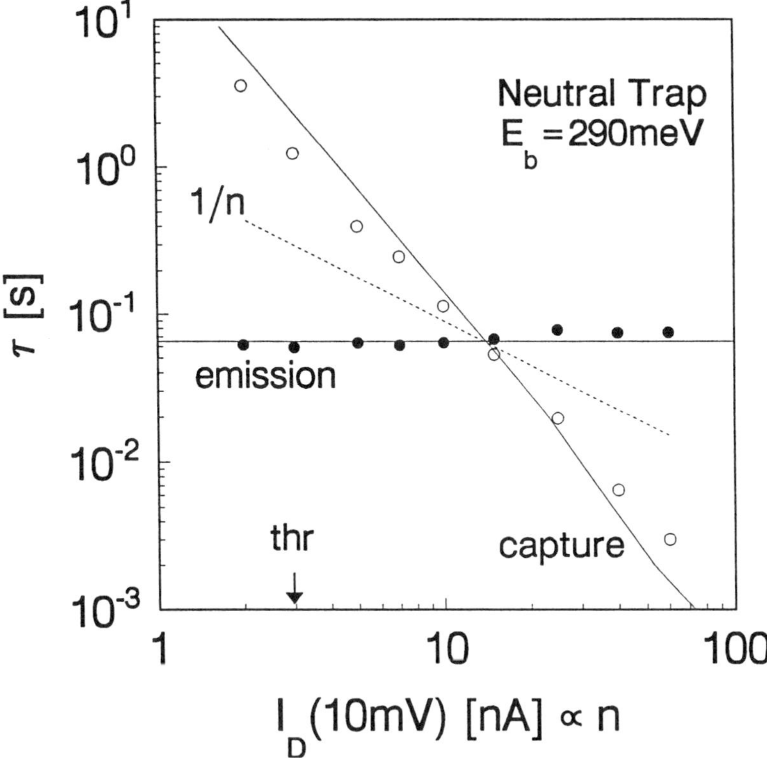

Figure 5.9: Capture and emission time constants of an individual neutral interface trap measured as a function of the dc source-drain current I_D at $V_D = 10$ mV ($n_s = 2 \times 10^9$ $I_D/$nAcm^2) by RTS (data points). The solid curves are calculated for a neutral trap with the Coulomb energy $\Delta E(n_s)$. Only two parameters ($E_b = 290$ mV and $C_n = 8 \times 10^{-14}$ cm^{3}s^{-1}) are used to fit the magnitude of the time constants.

Attractive Trap

The charge state of this trap is attractively charged (positive for electrons, negative for holes) when unoccupied. The trap becomes neutral when a mobile charge carrier is trapped from the channel. The typical dependence of the emission and capture times is depicted in Figure 5.10.

The capture time decreases with carrier density in the channel at low values of carrier density and turns to increase at high carrier desnities. The turnaround may not always be observable. The capture time is also frequently observed to be constant in a limited region around its minimum. The emission time steeply increases with carrier density sometimes over several orders of magnitude. This trap type is frequently observed, but is not as prevalent as the neutral trap.

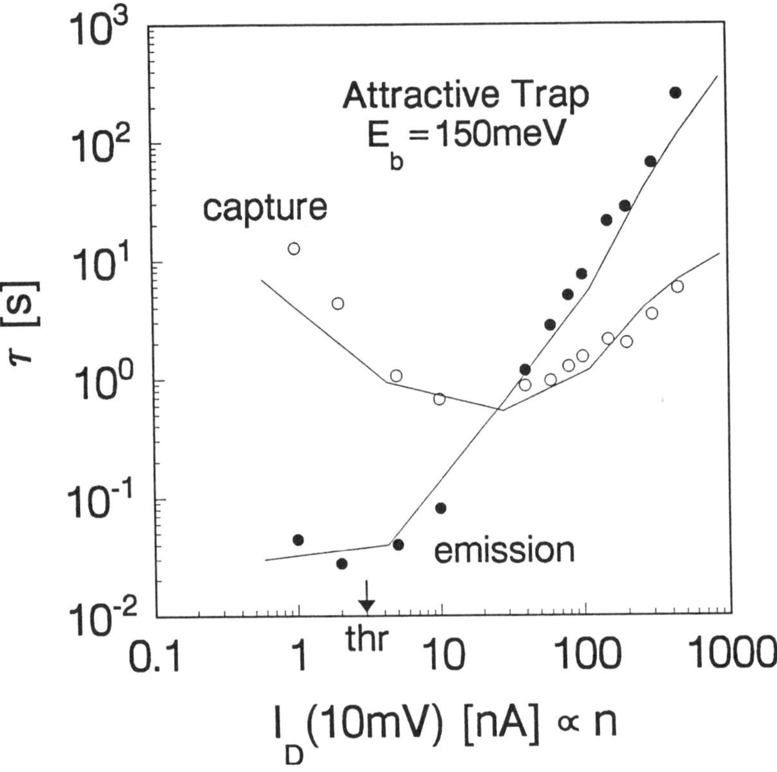

Figure 5.10: Capture and emission time constants of an individual attractive interface trap measured as a function of the dc source-drain current I_D at $V_D = 10$ mV ($n_s = 2 \times 10^9$ $I_D/$nAcm^2) by RTS (data points). The solid curves are calculated for an attractive trap with the Coulomb energy $\Delta E(n_s)$. Only two parameters ($E_b = 150$ mV and $C_n = 2 \times 10^{-18}$ cm^{3}s^{-1}) are used to fit the magnitude of the time constants.

Bistability

A bistable trap was first reported by Schulz and Karmann [7] at elevated temperatures ranging from 90 to 250 K and recently also by Cobden and Uren [16] at liquid He temperature. The bistable trap is always occupied by a charge carrier, which can assume two alternative configurations. The switching of the channel conductance is induced by a mobility change only. In a strict sense the definition of an emission or capture process is not useful, because the charge carrier is always bound to the trap; however, similar to all RTSs emission and capture times are attributed by the definition that the ratio of the two time constants must increase with carrier density in the channel and electric field in the oxide, *i.e.* the occupation factor of the trap must increase from 0 to 1. The

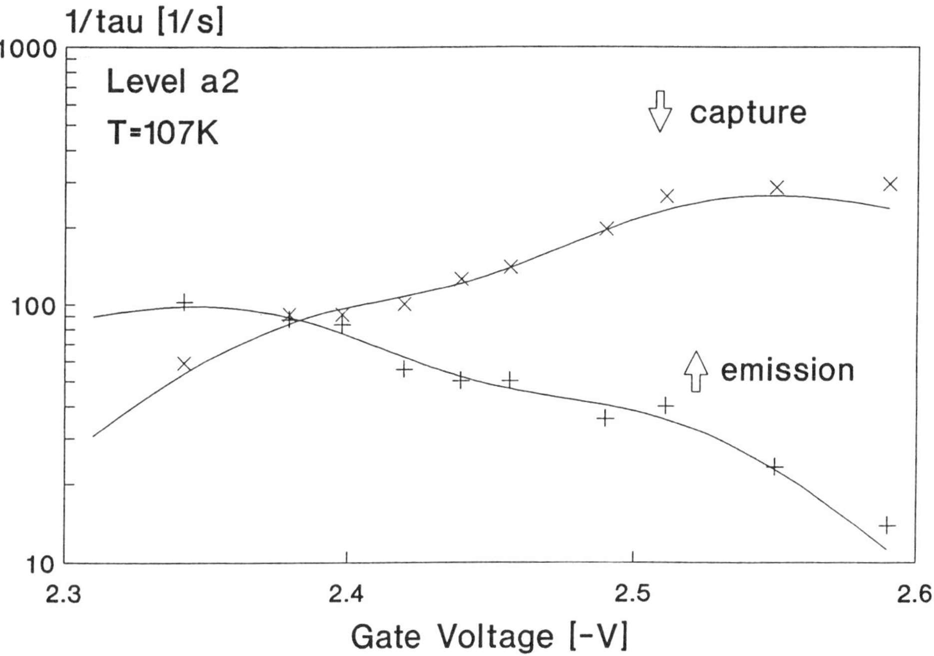

Figure 5.11: Capture and emission time constants of an individual bistable interface trap measured as function of the gate voltage. The curves are calculated for a two-level system.

switching may be anomalous, *i.e.* the high and low conductance states may be inversed with respect to the definition in Figure 5.1(b). For a bistability, both the emission time and the capture time vary with carrier density or gate bias more or less symmetrically due to a field dependence of the energy splitting of the two states. A typical dependence of the transition rates is shown in Figure 5.11.

The characteristic features are a negligible temperature activation because of tunneling and a symmetric intersection of the capture and emission time constants with constant product $\tau_e.\tau_c = \text{constant}$. Narrow tunneling resonances are superimposed especially at low temperatures [16]. The bistability is most frequently observed at low temperatures and seems to preferably occur after stressing of the device [16].

5.3.2 Temperature Activation

The evaluation of RTSs also delivers information about the thermal activation of the emission and capture rates of a single defect. An example is given in Figure 5.12 for a neutral/repulsive trap, where the rate constants normalized

142

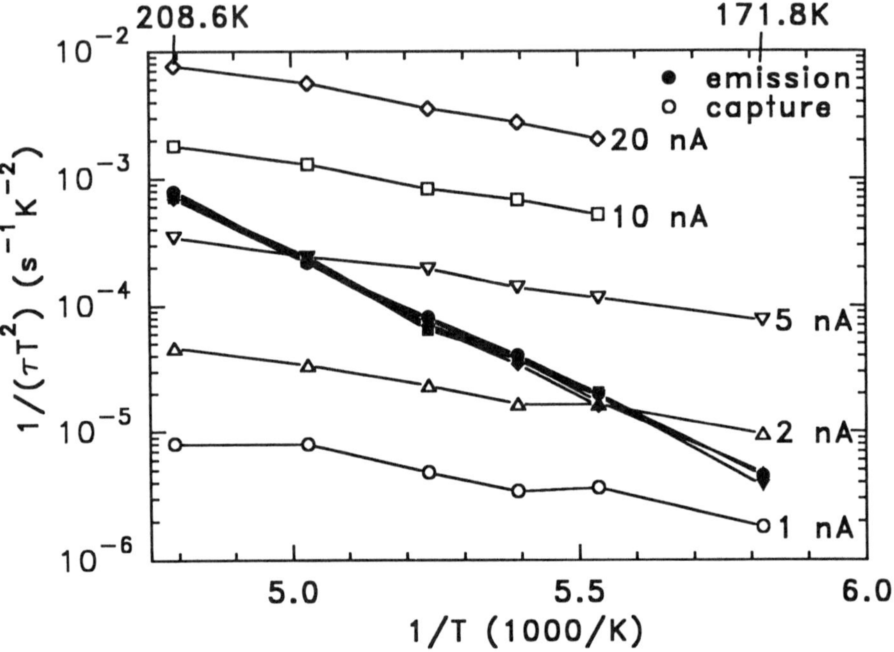

Figure 5.12: Arrhenius plot of an individual interface trap. The capture and emission rates are plotted as a function of inverse temperature. Parameter is the drain current as indicated at $V_D = 10$ mV.

to T^2 are plotted as a function of inverse temperature, as common for traps in bulk semiconductors. The thermal activation can be extracted from the slope in this Arrhenius plot. The activation energy is always higher for emission than for capture.

5.3.3 Location of Individual Traps

The individual trap levels causing RTSs must be located in the vicinity of the MOSFET channel in order to cause the switching of the conductance. Most likely defect sites are at the $Si - SiO_2$ interface or within a tunneling distance into the oxide in a defect-rich interfacial transition region of the oxide. If the defect is located at a distance into the oxide, the mobile carriers in the channel and the trap are spatially separated. The potential gradient across this distance leads to a trap level lowering with electric field in the oxide. Since the carrier density in the channel and this trap energy lowering are both dependent on the electric field, and the capture probability is dependent on both, the two effects cannot be separated in measurements of the gate bias dependence only.

The depth of traps in the oxide can thus be estimated by varying the gate bias and keeping the carrier concentration in the channel constant by applying a reverse substrate bias to the transistors. Thus the field across the oxide can be varied independent of the gate bias, and a possible field lowering of a trap in the oxide can be measured. The depth in the oxide can be approximately determined according to the relation [17]

$$d_T = \frac{kT t_{ox}}{q \Delta V_G} \left[\ln(\tau_e/\tau_c)_{V_{sub}} - \ln(\tau_e/\tau_c)_0 \right], \qquad (5.4)$$

where t_{ox} is the oxide thickness and ΔV_G the difference in gate voltage applied to yield the same carrier concentration for the cases with and without substrate bias. The presence of a channel is required for the application of this relation in order to pin the surface potential at the constant value. It is noted that this relation can only yield an estimation for the trap position, since the surface potential will change slightly when a substrate bias is applied and when due to the increased field the potential well at the interface gets narrower. Most of the traps evaluated turned out to be less than one nanometer away from the interface into the oxide. The field lowering caused by a trap location somewhat in the oxide is small compared to the contribution of the Coulomb energy to the trap energy [18] to be discussed below.

The location of the traps along the channel can be derived from interchanging the source and drain connections of the transistor in the measurement circuit and changing the sign of the source or drain voltage with respect to the ground. The change in trap occupation observed then corresponds to an effective change in the trap level energy and amounts to a fraction of the total source-drain voltage applied. This can be interpreted in terms of distance of the trap from source and drain [19].

5.3.4 The Modulation of the Channel Conductance

The conductance modulation ΔG of the channel is commonly made visible as a voltage or current amplitude with a constant current or a constant voltage applied, respectively. Its magnitude was discussed by Kirton *et al* [10] and by Ohata *et al* [14]. The modulation can be viewed as due to three effects, the number fluctuation of charge carriers in the channel, the change in carrier mobility caused by the creation or annihilation of a scattering center in the channel, and a geometrical change of current flow in a nonuniform channel. Averaging the effects over the entire device yields

$$\Delta G = \Delta(f_g q \mu n_s) = f_g q \mu . \Delta n_s + f_g q n_s . \Delta \mu + q \mu n_s . \Delta f_g \qquad (5.5)$$

where q is the electronic charge, μ the mobility of channel charge carriers and $\Delta \mu$ its effective change averaged over the channel, n_s is the concentration of inversion charge per unit area and Δn_s its change due to the trapping event. The parameter f_g is a geometrical factor for non-uniform current flow. For a uniform channel f_g equals W/L, W and L being the channel width and length,

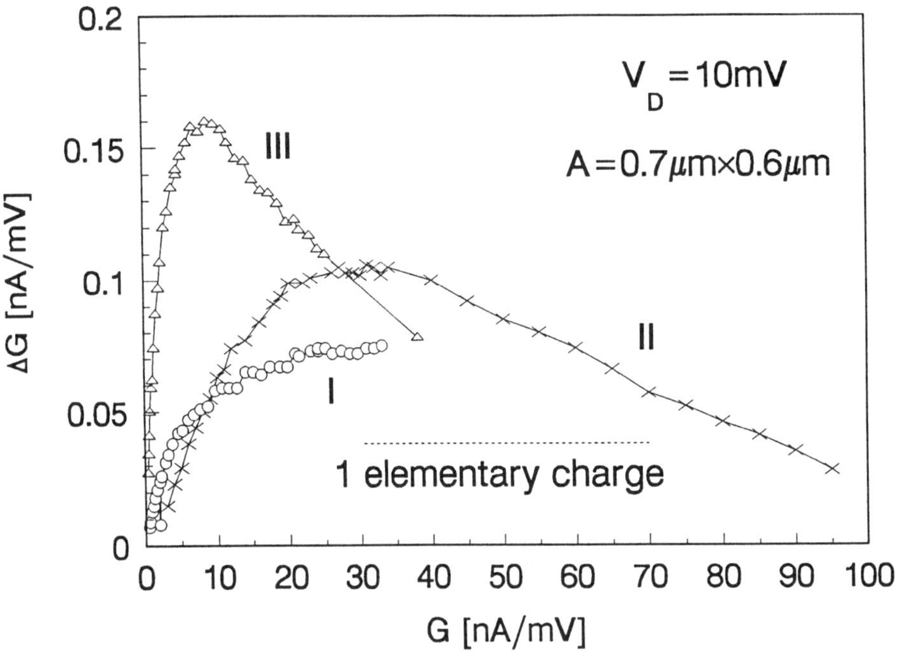

Figure 5.13: Switching conductance step ΔG of three defects from a small-area MOSFET vs. channel conductance G. The numbers I-III refer to the same individual traps as those in Figure 5.6. The driving and sample parameters are: $V_D = 10$ mV, T $= 300$ K, device dimensions: 0.6μm $\times$ 0.7μm.

respectively. The variation of the factor f_g expresses the effect of a modulation of the local current density due to the perturbation induced by the trap. Such effects are known to occur near the drain, especially after stressing [20]. These have so far not been considered in the magnitude of RTSs. In a non-uniform channel, RTSs with large amplitudes are found with relative ΔG/G values of as much as 30 percent, which cannot be solely explained by a number variation of inversion charge carriers. They may be due to a strong mobility effect, in combination with a redistribution of the current flow in the whole device due to a possible strategic location of the trap in a confined current path under the gate area.

When the conductance step ΔG caused by the switching of a single trap is plotted vs. the channel conductance, the typical behavior depicted in Figure 5.13 is found.

The perturbation induced by a single trapped charge initially increases with increasing channel conductance, assumes a maximum, and decays again for high carrier density in the channel in strong inversion. The initial increase is mainly due to the increase of mobile charge carriers in the channel. In strong inversion the charge trapped is screened by the channel charge carriers, so that the change in conductance there decreases and finally approaches that for one carrier added or removed. The changes induced in general also depend on the location of the trap in the channel and on potential fluctuations induced by fixed charges in the oxide.

5.4 Interpretation and Modeling

5.4.1 Coulomb Energy

The major time constant effects of RTSs could recently be explained by the free-energy change of the Coulomb energy induced by the transfer and localization of a single electron charging the capacitor electrodes [21]. The comparison of this model with RTS was made first on the basis of published data [17] and verified by a set of RTSs observed in different MOSFETs [22, 18]. The Coulomb effect is known in the literature as the Coulomb blockade [23-27]. It is usually discussed for charging a small, floating capacitor with a single electron by tunneling. The magnitude of the Coulomb blockade is usually only a few millivolts. In MOS structures, high Coulomb effects are possible in the transfer and localization of a single electron into an interface trap.

The energies involved in the transfer of an electron to and its capture into an individual interface trap located at the $Si - SiO_2$ boundary of an MOS structure are depicted in Figure 5.14.

The band diagram in Figure 5.14 is for a p-type semiconductor with the MOS structure biased in weak inversion. An electron (-q) to be trapped in the interface may be viewed as taken from the Fermi level outside the space charge region (equilibrium level) and transferred to the interface into the interface trap level at energy E_{T0}. In order to maintain charge neutrality, fractional counter charges of the elementary charge (image charges) are created on the electrodes of the capacitor structure, *i.e.* on the gate (Q_G), in the semiconductor substrate (Q_S) and, if present, in the inversion channel (Q_{ch}).

The transfer of an electron to the interface requires a Coulomb energy to charge up the capacitor [21]. For the non-linear capacitance of the semiconductor space charge region, this energy is not completely represented by the potential energy as depicted by the band bending in Figure 5.14. The Coulomb energy ΔE of an electron captured into an interface trap may be calculated from the balance of the energy required to charge the capacitor electrodes and that of the transfer of a charge carrier to the interface into the trap level.

$$
\begin{aligned}
\Delta E &= \frac{Q_G}{q}(E_F - E_{FM}) + \frac{Q_S}{q}(E_F - E_{V0}) + \frac{Q_{ch}}{q}(E_F - E_C) - (E_F - E_{T0}) + E_b \\
&= \frac{Q_G}{q}(E_F - E_{FM}) + \frac{Q_S}{q}(E_F - E_{V0}) + \frac{Q_{ch}}{q}(E_F - E_C) - (E_F - E_C)
\end{aligned}
\tag{5.6}
$$

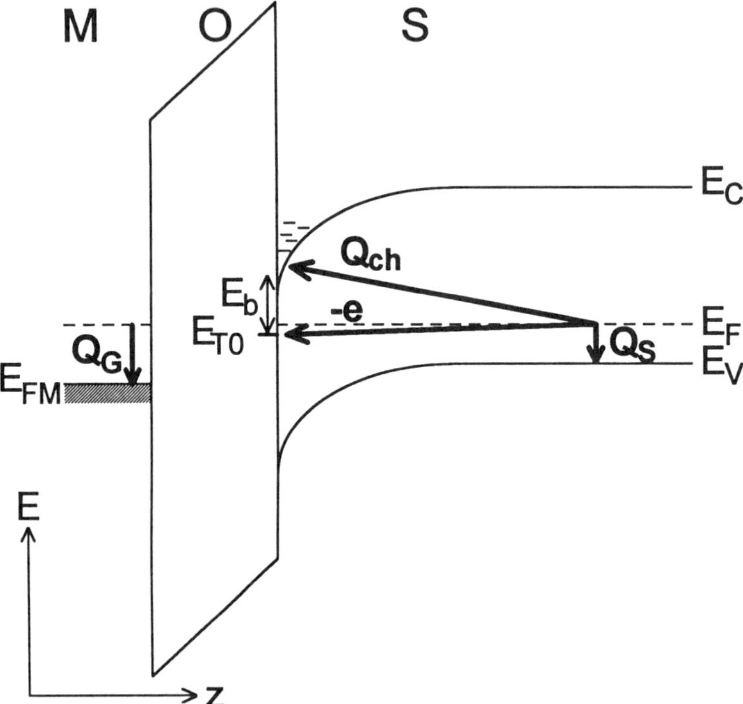

Figure 5.14: Band diagram of an MOS structure biased into inversion. An electron charge –q is transferred from the semiconductor Fermi energy E_F into the interface trap level E_{T0} located by the binding energy E_b below the conduction band edge. Fractional counter charges +Q have to be created from E_F as indicated by the arrows to form image charges on the gate, in the channel and at the boundary of the space charge region.

In order to remove the unknown binding energy $E_b = E_C - E_{T0}$ of the electron to the trap, the constant term E_b is added to the energy balance of the transfer. This correction is equivalent to moving all the differing traps up to the band edge E_C (second line of Equation 5.6), thus making the Coulomb energy independent of trap-specific parameters. In Equation 5.6, E_F and E_{FM} are the Fermi energies in the semiconductor and in the gate metal, respectively, E_{V0} is the energy of the valence band edge in the semiconductor substrate, and E_C the conduction band edge at the interface.

The image charges on the gate Q_G, in the substrate Q_S, and in the channel Q_{ch}, and the Coulomb energy ΔE have been numerically calculated by a finite element method solving Poisson's equation of the MOSFET capacitor structure with an elementary point charge at the $Si - SiO_2$ interface. The partial charges Q and the Coulomb energy ΔE are depicted in Figures 5.15 and 5.16, respectively, as a function of the gate bias voltage. The gate bias drives the MOS

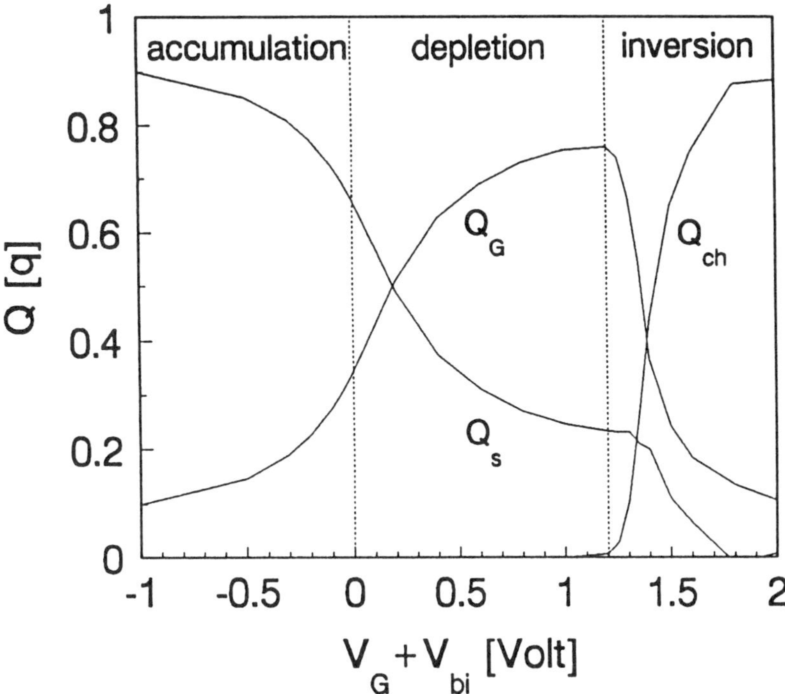

Figure 5.15: Calculated image charges created by the transfer of a single electron into an interface trap at T=295 K. Q_G, Q_S and Q_{ch} are the fractions of a positive elementary charge on the gate, in the substrate and in the channel, respectively. The MOSFET parameters are the oxide thickness $t_{ox} = 10$ nm and semiconductor substrate doping $N_A = 10^{23}$ cm^{-3}. The gate bias voltage ranges from accumulation (left) via depletion (middle) to inversion (right).

capacitor structure from accumulation via depletion into inversion. The band bending and the width of the space charge region are functions of the sum of the applied gate bias V_G and of the constant built-in voltage V_{bi}, which is due to e.g. the workfunction difference between the metal and the semiconductor. The image charges depend only on the sum of the two voltages. The Coulomb energy is a function of both, the applied bias voltage and the built-in voltage.

Figure 5.15 depicts the image charges Q_G, Q_S, Q_{ch} as calculated at temperature $T = 295$ K for an n-channel MOS capacitor structure having a gate oxide thickness of $t_{ox} = 10$ nm and a semiconductor substrate doping concentration of $N_A = 10^{23}$ m^{-3}.

The fractional image charges add up to one elementary counter charge $+q$ of the electron captured into the interface trap. In accumulation only a small fraction of image charge is found on the gate because of the partial screening

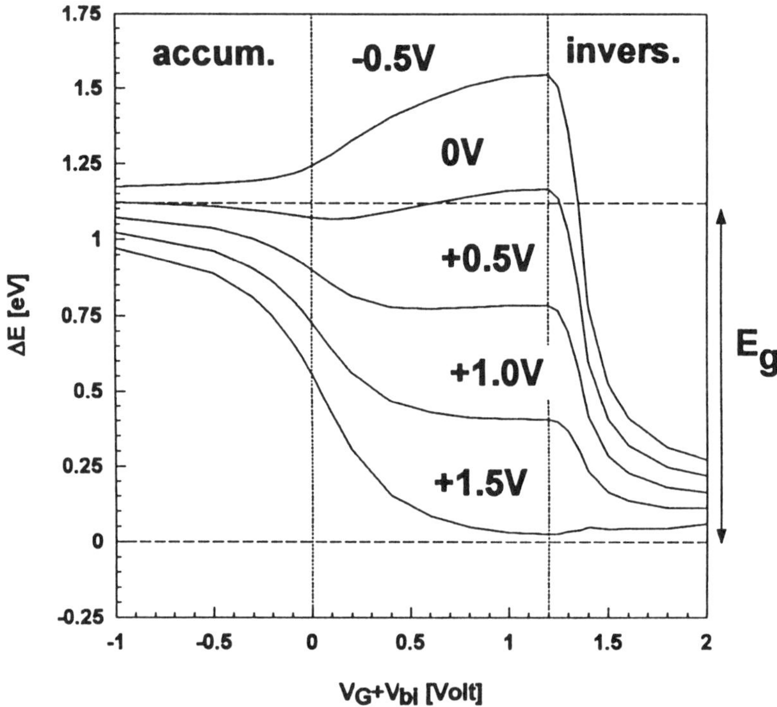

Figure 5.16: Coulomb energy ΔE vs. gate voltage calculated for an n-channel MOSFET at T=295 K having the same data as in Figure 5.15. Parameter is the built-in voltage V_{bi} of the device. The built-in voltage for the experimental data in Figure 5.17 is $V_{bi} = 1.0$ V.

of the trapped electron by the mobile majority carriers in the semiconductor substrate. The charge on the gate increases when in depletion the space charge region widens with increasing gate bias. With the onset of inversion, both the gate charge and the substrate charge tend to disappear when the mobile minority charge carriers in the channel screen the trapped charge. The weak screening of the two-dimensional channel [28] causes the gradual decay of the image charges [29].

The Coulomb energy is depicted in Figure 5.16. It is a function of the gate bias voltage V_G and of the built-in voltage V_{bi}, which is given as a parameter. The built-in voltage is $V_{bi} = 1.0$ V for some experimental data presented later in this chapter. In strong inversion the Coulomb energy slowly approaches zero because the image charge on the gate is strongly screened by the mobile minority charge carriers in the channel. In the accumulation region the image charge on the gate is strongly screened by the mobile majority carriers. The Coulomb energy in this case approaches the band gap energy $E_g = 1.1$ eV, because the trap level is referenced from the conduction band ($E_b = E_C - E_{T0}$) and in accumulation the binding energy for minority carriers is shifted by E_g. In the depletion regime the Coulomb energy increases weakly with gate bias and reaches a maximum before, at the onset of inversion, the screening by minority carriers becomes effective and reduces the Coulomb energy towards zero.

Since RTS measurements of the Coulomb energy require a conductive channel, an analysis is only possible in the inversion regime of Figure 5.16. In this regime, the Coulomb energy may assume large values for weak channels. It gradually decays for strong inversion. It is noted that the Coulomb energy is large for small or negative built-in voltages (n-channel) and is small for built-in voltages in excess of the band gap energy, *i.e.* for normally-on transistors.

The Coulomb energy may assume large values of several hundred millivolts under bias conditions in depletion, where the image charge on the gate is also large. The Coulomb energy gradually disappears with screening by mobile charge carriers in the inversion channel of MOSFETs. The Coulomb energy is therefore a strong function of the carrier density in the MOSFET channel. The effect of the Coulomb energy varies for the different trap types classified in section 5.3.1.

A- Neutral Trap

A neutral trap center in the MOSFET is charged repulsively when occupied. In this case the Coulomb energy ΔE additionally activates the capture rate, since a carrier transferred to the location of the trap has to supply the Coulomb energy. The emission rate remains unaffected, because the effective barrier is the binding energy. The rate equations for capture and emission are obtained by implementing the SRH laws of detailed balance with the Coulomb energy ΔE as

$$\frac{1}{\tau_c} = C_n N_C \exp\left[-(E_C - E_F)/kT\right] . \exp\left(-\Delta E/kT\right) \tag{5.7}$$

and

$$\frac{1}{\tau_e} = C_n N_C \exp\left(-E_b/kT\right) \tag{5.8}$$

The equilibrium occupation factor of the trap level is now affected by the Coulomb energy ΔE:

$$\frac{\tau_e}{\tau_c} = \exp\left[-(E_C - E_b - E_F + \Delta E)/kT\right] \tag{5.9}$$

The total trap energy E_T effective in the equilibrium occupation factor is raised by the Coulomb energy ΔE, which is dependent on channel carrier density and temperature, from $E_C - E_b$ to $E_C - E_b + \Delta E(n_s, T)$.

The experimental results presented in Figure 5.9 for the neutral trap were compared with the theoretical modeling of Equation 5.6 for the Coulomb energy [21]. The image charges and the Coulomb energy were calculated solving

150

Poisson's equation with a finite element method [29]. The calculations are depicted in Figure 5.9 by the continuous lines. The constant trap binding energy and a constant capture coefficient are fitting parameters defined by the intersection point of the emission and capture time curves. The shape of the curves, especially the super-linear dependence of the capture time constant is well described by the modeling.

B- Attractive Trap

For trapping into an attractive defect center the Coulomb charge is already present and is neutralized by the charge carrier to be trapped. The electron gains Coulomb energy by the transfer towards the trap location. In this case, both the capture and the emission rates are affected by the Coulomb energy. The resulting rates are for the capture an enhancement by the Coulomb energy

$$\frac{1}{\tau_c} = C_n N_C \exp\left[-(E_C - E_F)/kT\right] . \exp\left(+\Delta E/kT\right) \tag{5.10}$$

and for the emission also an enhancement

$$\frac{1}{\tau_e} = C_n N_C \exp\left(-E_b/kT\right) . \exp\left(+\Delta E/kT\right) \tag{5.11}$$

The equilibrium occupation factor remains unaffected by the Coulomb energy

$$\frac{\tau_e}{\tau_c} = \exp\left[-(E_C - E_F - E_b)/kT\right] \tag{5.12}$$

The trap energy E_T effective in the equilibrium occupation of the attractive center is the constant binding energy similar to trapping in a bulk center.

The experimental results of the attractive center presented in Figure 5.10 were also compared with the modeling of the Coulomb energy. The calculations are depicted in Figure 5.10 by the continuous lines. Both the emission and the capture times are strongly affected by the Coulomb energy, which is large ($\Delta E > 230$ meV) for low channel charge carrier densities and which reduces to small values ($\Delta E \approx 40$ meV) for high carrier densities. The increase of the emission time (4 orders of magnitude) directly represents the diminishing Coulomb energy. The turnaround of the capture time is due to the decaying Coulomb energy by the increasing Fermi energy in Equation 5.10 with increasing carrier density. The calculated curves represent the measurement data very well. Only the constant binding energy and the constant capture coefficient are data fitting parameters.

5.4.2   Unified Coulomb Energy

For the comparison of the Coulomb energy with the experimental data in Figures 5.9 and 5.10, two fitting parameters were required, which obscured the fundamental property of the Coulomb energy. In References [22] and [18], Mueller

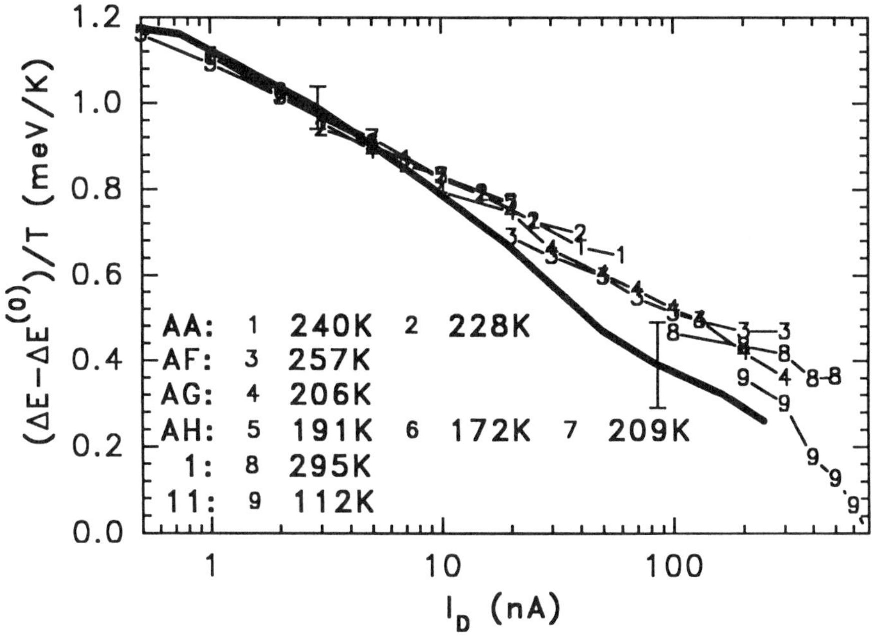

Figure 5.17: Temperature-normalized Coulomb free energy $(\Delta E - \Delta E^{(0)})/T$ as a function of drain current I_D at constant source-drain voltage $V_D = 10$ mV. $\Delta E^{(0)}$ vanishes for interface traps referenced to the band edge. The numbers indicate the RTSs as given in table 5.1. The bold line is calculated by the theoretical model with the transistor data.

et al demonstrate an evaluation method to determine the binding energy and the capture coefficients of individual traps from RTS measurement data as a function of temperature and channel carrier density. The Coulomb energy for neutral traps may then be determined from Equation 5.7 as the remaining activation energy after correction with the binding energy and the calculated Fermi energy. It is found that the Coulomb energy is proportional to the temperature. A unified curve is obtained for all the individual trap levels studied by RTS at different temperatures and in different MOSFETs, when the Coulomb energy normalized to temperature $\Delta E/T$ is plotted, as a function of channel carrier density. This is demonstrated in Figure 5.17. All the different traps line up onto one curve, which is in good agreement with the theoretical modeling of Equation 5.6, represented by the thick curve in Figure 5.17, *without any fitting parameter*. The theoretical curve is taken from Figure 5.16 for $V_{bi} = +1$ V.

The calculations of the image charges and of the Coulomb energy can so far only be performed numerically. An explicit functional dependence is not known. An empirical relation is deduced from the experimental data. For the

MOST (#)	RTS (#)	T (K)	$E_b - \Delta E^{(0)}$ (meV)	$C_n = \sigma_n v_{th}$ ($cm^3 s^{-1}$)	σ_{n0} (cm^2)	E_σ (meV)
AA	2	228	280	5×10^{-14}	9×10^{-14}	340
	1	240		1×10^{-13}		
AF	3	257	180	2×10^{-15}	4×10^{-13}	490
AG	4	206	175	1×10^{-13}	1×10^{-14}	255
AH	6	172	270	9×10^{-13}	5×10^{-15}	170
	5	191		3×10^{-12}		
	7	209		7×10^{-12}		
1	8	295	210	2×10^{-14}	4×10^{-15}	390
11	9	112	20	2×10^{-17}	2×10^{-16}	180

Table 5.1: List of trap data evaluated from several RTSs. Listed are the sample (MOSFET) identification (column 1), the RTS number (column 2), measurement temperature T (column 3), constant trap binding energy E_b reduced by a possible constant offset $\Delta E^{(0)}$ of the Coulomb energy (column 4), capture coefficient C_n (column 5), capture cross section prefactor σ_{n0} (column 6) and its temperature activation E_σ (column 7).

MOSFETs having oxide thicknesses of $t_{ox} = 10$ nm, $V_{bi} = +1$ V, the data are well described in the interval between $I_D = 1$ nA and 100 nA by

$$\Delta E - \Delta E^{(0)} = -aT \ln \left(I_D / I_0 \right) \tag{5.13}$$

with a $= 0.129$ meV/K and $I_0 = 5.37$ μA. Equation 5.13 has striking similarity to an entropy change. An entropy contribution to RTS trapping rates has been discovered earlier in the experimental data and attributed to lattice properties [4]. The quantitative agreement with the Coulomb energy, however, indicates that the entropy originates predominantly from the reconfiguration of mobile charge carriers.

The trap properties determined are compiled in Table 5.1. For the randomly selected trap centers, also a random scatter of the binding energy values is observed. The binding energy is in all cases less than half the band gap and thus consistent with minority carrier trapping. The capture coefficient is typically several orders of magnitude smaller than for bulk trapping due to its additional activation energy E_σ. The deviation from the classical SRH laws observed for the capture and emission rates of interface traps is predominantly caused by the Coulomb energy. Additional entropy or electric field effects on interface trapping are marginal ($T\Delta S < kT$) as compared to the Coulomb effect, which quantitatively describes the magnitude observed.

5.4.3 Activation of the Capture Coefficient

The RTS time constants reveal the presence of a considerable additional barrier E_σ effective in the capture coefficient (Table 5.1). This activation barrier amounts to several hundred meV and is specific to individual trap centers. The barrier can be clearly separated from the Coulomb energy and the oxide field effect. Various models have been proposed to explain this phenomenon. Kirton and Uren [6] suggest a trap distortion with multiphonon relaxation as first proposed for bulk traps [30]. Another origin may be an interfacial barrier at the $Si - SiO_2$ interface, which the carriers have to surmount [31]. Both effects are likely to occur in the MOS structure when the traps are located somewhat in the amorphous oxide film. Both models cannot be separated by the experimental dependencies of the time constants observed in RTSs.

5.5 Conclusion

In small-area MOSFETs, the alternate capture and emission of carriers into and from single, individual defects in the oxide generates discrete switching in the source-drain resistance. The resistance changes are observed in the device current as a random telegraph signal or as a quantized transient after a trap filling pulse.

Noise measurements on small MOSFETs demonstrate that the 1/f noise power spectrum is a superposition of several Lorentzians which originate from random telegraph signals of individual traps covering a wide spectrum of time constants.

All the traps exhibit a universal Coulomb energy which is proportional to temperature and which decreases with inversion carrier density due to screening. The trapping processes are classically modeled in analogy to SRH statistics, taking into account the Coulomb energy. The traps are located within 1 nm from the interface into the oxide. Trap entropy effects are small, of the order of 1k. Additional activation barriers of several hundred meV are observed in the capture coefficients. The chemical nature of the defects causing the RTSs is still unknown.

It seems that the MOS capacitance-voltage (CV) and deep level transient spectroscopy (DLTS) measurements and evaluations have to be revised to take into account the large Coulomb energies involved in interface trapping as revealed by RTS measurements [21].

Acknowledgment

The authors gratefully acknowledge the continuous support of work in this field by the Siemens Corporate Research Lab, Munich.

Bibliography

[1] Farmer, K. R., C. T. Rogers, and R. A. Buhrmann, *Phys. Rev. Lett.*, 58, 2255 (1987).

[2] Welland, M. B., and R. H. Koch, *Appl. Phys. Lett.*, 48, 724 (1986).

[3] Peng, L. L., P. C. Canfield and D. A. Allstot, *IEEE Trans. Electron Devices*, ED-39, 2444 (1992).

[4] Kirton,. M. J., and M. J. Uren, *Advances in Physics*, 38, No. 4, 367 (1989).

[5] Ralls, K. S., W. J. Skocpol, L. D. Jackel, R. E. Howard, L. A. Fetter, R. W. Epworth and D. M. Tennant, Phys. Rev. Lett., 52, 228 (1984).

[6] Kirton, M. J., and M. J. Uren, *Appl. Phys. Lett.*, 48, 1270 (1986).

[7] Schulz, M., and A. Karmann, *Appl. Phys. A*, 52, 104 (1991); cf also *Proceedings of the International Conference on Insulating Films on Semiconductors INFOS'91*, eds. W. Eccleston and M. J. Uren, Adam Hilger, Bristol, 143 (1991).

[8] Schulz, M., and A. Karmann, *Physica Scripta*, T35, 273 (1991).

[9] Farmer, K. R., *Insulating Films on Semiconductors, INFOS'91*, eds. W. Eccleston and M. J. Uren, Adam Hilger, Bristol, 1 (1991).

[10] Kirton, M. J., M. J. Uren, S. Collins, M. Schulz, A. Karmann and K. Scheffer, *Semicond. Sci. Technol.*, 4, 1116 (1989).

[11] Karwarth, A., and M. Schulz, *Appl. Phys. Lett.*, 52, 634 (1988).

[12] Papoulis, A., *Probability, Random Variables, and Stochastic Processes*, McGraw-Hill, New York (1965).

[13] Schulz, M. and A. Pappas, in *Noise in Physical Systems and 1/f Fluctuations*, T. Musha, S. Sato and M. Yamamoto eds., Ohmsha Ltd., Tokio, 265 (1991).

[14] Ohata, A., A. Toriumi, M. Iwase and K. Natori, *J. Appl. Phys.*, 68, 200 (1990).

[15] Shockley, W., and W. T. Read, jr., *Phys. Rev.*, 87, 835 (1952).

[16] Cobden, D. H., and M. J. Uren, in *Proceedings of the International Conference on Insulating Films on Semiconductors INFOS'93*, Delft, The Netherlands, P. Balk ed., Elsevier Science Publishers, Amsterdam, 163 (1993).

[17] Schulz, M., A. Pappas and J. Vennemann, in *The Physics and Chemistry of SiO_2 and its Interfaces to Silicon*, C. R. Helms and B. E. Deal eds., Plenum Press, New York, 383 (1993).

[18] Mueller, H. H., D. Wörle and M. Schulz, *J. Appl. Phys.* (1994).

[19] Restle, P., *IBM J. Res. Dev.*, 34, 227 (1990).

[20] Bollu, M., F. Koch, A. Madenach and J. Scholz, *Appl. Surf. Sci.*, 30, 142 (1987); also in *Proceedings of the International Conference on Insulating Films on Semiconductors INFOS'87*, eds. G. Declerck and R. de Keersmaecker, North Holland, Amsterdam, 142 (1987).

[21] Schulz, M., *J. Appl. Phys.*, 74(4), 2649 (1993); cf also in *Proceedings of the International Conference on Insulating Films on Semiconductors INFOS'93*, Delft, The Netherlands, P. Balk ed., Elsevier Science Publishers, Amsterdam, 171 (1993).

[22] Mueller, H. H., D. Wörle and M. Schulz, *Microelectronic Engineering*, 22, 181 (1993); also in *Proceedings of the International Conference on Insulating Films on Semiconductors INFOS'93*, Delft, The Netherlands, P. Balk ed., Elsevier Science Publishers, Amsterdam, 181 (1993).

[23] Likharev, K. K., *IBM J. Res. Dev.*, 32, 144 (1988).

[24] Chou, S. Y., and Y. Wang, *Appl. Phys. Lett.*, 61, 1591 (1992).

[25] Schöneberger,C., H. van Houten and H. C. Donkersloot, *Europhysics Lett.*, 20, 249 (1992).

[26] Berthe, R., and J. Halbritter, *Phys. Rev. Lett.*, 43, 6 (1991).

[27] *Single Charge Tunneling: Coulomb Blockade Phenomena in Nanostructures, Proceedings of NATO Advanced Study Institute on Single Charge Tunneling*, H. Grabert ed., Plenum Press, New York (1992).

[28] Ando, T., A. B. Fowler and F. Stern, *Rev. Mod. Phys.*, 54, 437 (1982).

[29] Schulz, M., *Proceedings of Insulating Films on Semiconductors INFOS'91*, eds. W. Eccleston and M. J. Uren, Adam Hilger, Bristol, 127 (1991).

[30] Henry, C., and D. V. Lang, *Phys. Rev. B*, 15 989 (1977).

[31] Karmann, A., and M. Schulz in *Proc. of the Int. Conf. on Insulating Films on Semiconductors INFOS'89*, F. Koch and A. Spitzer eds., North-Holland,Amsterdam, 500 (1989).

6

Characterization of SOI MOSFETs

Sorin Cristoloveanu
Laboratoire de Physique des Composants à Semiconducteurs
Grenoble, France

6.1 Introduction

After a long period of maturation, SOI technology has reached a stage of rapid
and successful development. A major condition for competing with bulk silicon
in commercial applications is the degree of understanding of material properties, fabrication techniques, and device operation. This implies the use of
adequate characterization methods which overcome the difficulties induced by
the thinness of the film and by the stacked interfaces. Since the quality of SOI
structures has considerable impact on the performance of integrated circuits,
reciprocally the devices may be interrogated on the properties of the starting
material and the process–induced modifications.

In this chapter, the discussion of electrical characterization techniques will
proceed systematically from the wafer to the device. The *pseudo*–MOS transistor or Ψ–MOSFET is appropriate for *in-situ* inspection of the quality of the
buried oxide and interfaces in as-grown SOI wafers. More or less conventional
transport measurements provide the carrier concentration and mobility in the
thin silicon film.

The MOS capacitance and conductance techniques are part of the history of bulk silicon. However, these methods are far more difficult and fragile in SOI, where both the measurement and the parameter extraction are complicated by the contributions of several interfaces. Current-based techniques are more suitable for thin films: the current is easily measurable even in small area devices, the experiment can be controlled from the front or back gate, and simple single-interface models often are sufficient for data analysis. The usefulness of diodes and transistors for characterization purposes will be demonstrated by focusing on MOS–Hall profiling, charge pumping, low-frequency noise spectroscopy, Zerbst-like drain current transients, etc. A brief survey of the experimental set-up, appropriate models, and parameter extraction will be made for each technique. The relevant properties of SOI materials that are inferred from the characterization procedure will be discussed in terms of carrier generation lifetime and mobility, front and back interface traps, film and oxide defects.

6.2   Interest of MOS–SOI Technology

In an MOS transistor, only the very top region (0.1–0.2 μm thick) of the silicon wafer is actually relevant for electron transport. SOI structures emerged from the idea of isolating the active device overlay from the detrimental influence of the silicon substrate. The thickness of SOI films can be precisely adjusted to meet the requirements of any type of devices. Reviewed below are the main advantages of SOI devices [1].

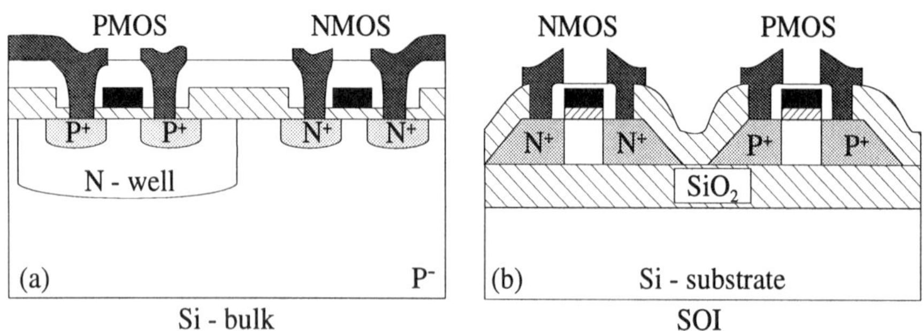

Figure 6.1: Schematic configuration of CMOS transistors in bulk silicon and SOI wafers.

Dielectric isolation.   SOI circuits consist of single-device islands, dielectrically isolated from each other and from the underlying substrate. The difference between SOI and bulk silicon is illustrated in Figure 6.1. The inter-device distance is much more shrinkable in SOI than in bulk. The critical limitation of bulk silicon for VLSI circuits comes from the unavoidable proximity of diffused regions that belong to adjacent components.

Sophisticated schemes of trench isolation are necessary in bulk silicon to avoid latch-up. By contrast, SOI is naturally free from latch-up.

Vertical junctions. In regular SOI films, the source and drain regions extend to the insulator and only their lateral sides serve as junctions. The surface of such a vertical junction is far smaller than in bulk silicon. This yields a substantial reduction in parasitic capacitances, hence in commutation delay and dynamic power dissipation. In other words, for a predefined power consumption, much denser and faster circuits can be integrated on SOI.

Circuit design & processing. The fabrication of CMOS circuits on silicon on insulator is simpler than in bulk silicon. The absence of wells and inter-device trenches reduces the number of processing steps and offers additional design flexibility. The location of individual device islands does not have to obey any isolation constraint, and is only aimed to minimize packing density and interconnections. The basic advantages of SOI — less processing, enhanced performance, more chips on the wafer — are key sources of profit. Economical evaluations are quite positive about turning the mass production of ULSI circuits from bulk silicon to SOI.

Short-channel effects. In small geometry MOS transistors, oppressing short-channel effects originate from 'charge sharing' between gate and junctions: threshold voltage roll-off (see also Figure 6.10(b)), degradation of the sub-threshold slope, drain induced barrier lowering, punchthrough, bipolar transistor action, etc. SOI devices are more immune to short-channel effects. Basically, the extension of source/drain depletion regions is restricted by the junction size and by the dual-gate control of the surface potentials.

Low voltage/power operation. SOI MOSFETs stand as the ideal candidate for the next generation of low power, low voltage integrated circuits. It is expected that the components for mobile communication will be designed for long-term one-battery operation (0.9–1.5 V bias supply). In thin fully-depleted SOI transistors, the subthreshold slope is sharper and the leakage currents are much lower than in bulk silicon MOSFETs. This offers the opportunity for reducing the threshold voltage around 0.3 V. In addition, the higher speed of SOI circuits and their better tolerance to short-channel effects will avoid a severe drop in the performance brought about by the drastic reduction of the bias supply.

Reliability. The primary motivation for developing SOI technologies was their excellent tolerance of *transient* radiation effects. Incoming particles generate electron-hole pairs in proportion to the device volume. Resulting photo-currents act as leakage currents causing charge collection and soft errors. Logic upsets are remarkably reduced in SOI since the volume exposed to carrier generation is 2–3 orders of magnitude smaller. Another reliability issue is the vulnerability of submicron devices to hot carrier injection. The basic argument, regarded as a promise for enhanced immunity to hot carrier damage, is the reduction of the electric field peak in SOI.

The enthusiastic enumeration of SOI merits and potential advantages over bulk silicon, was not enough to stop the incredible progress of bulk Si-based microelectronics. For each new generation of components, bulk silicon technology was able to overcome its limitations and to progress beyond the most optimistic expectations. It is this success which has primarily delayed the entering of SOI in the commercial arena.

Besides the health of bulk silicon technology, other factors have inhibited so far the development of SOI. The SOI substrates are not fully optimized in terms of uniformity, availability, and cost. An additional drawback and challenging issue of SOI is the control of specific effects in SOI transistors, such as floating body, dynamic transients and interface coupling.

6.3 Synthesis of SOI Structures

6.3.1 SIMOX

SIMOX is considered to be the most advanced and promising of SOI technologies for high density CMOS circuits. The buried oxide is synthesized by internal oxidation during the deep implantation of oxygen ions into silicon. High current implanters (100–200 mA) have been designed to produce large SIMOX wafers with excellent purity and homogeneity. The typical implantation conditions of commercially available SIMOX wafers are $1.8 \times 10^{18}\,O^+/cm^2$, 190 keV, and $650\,^\circ C$. The thicknesses of the Si overlay and buried oxide (BOX) are 210 nm and 380 nm, respectively.

A post-implantation annealing is necessary to recover the crystalline quality of the Si overlay. High temperature annealing, up to $1360\,^\circ C$, is performed in poly-Si furnaces [2]. The recrystallization is a nonlinear effect of annealing temperature, duration and temperature ramping. Although the ideal conditions are not fully identified yet, it is admitted that very good SIMOX needs annealing at about $1320\,^\circ C$, for 6 hours, in argon ambient containing 1% of oxygen.

A number of SIMOX variants have been explored (Figure 6.2): thin and thick films, ultra-thin BOX, interrupted oxide. The double SIMOX structure is fabricated by two sequential oxygen implants at high and low doses. The silicon layer sandwiched between the two oxides is useful for interconnects, electrical shielding, or optical waveguides. Totally isolated islands were processed by masked implantation through a patterned capping oxide. The implantation of nitrogen instead of oxygen results in sharper interfaces after LTA, but the level of donor autodoping is unacceptable. Mixed implantations of oxygen and nitrogen or oxygen and carbon were tested with the aims of hardening the BOX and attenuating the electrical activity of oxygen donors [3]. The best quality SIMOX material so far is prepared by a multi-step process composed of sequential implants of low doses (0.3–$0.6 \times 10^{18}\,\mathrm{O^+/cm^2}$) and intermittent HTA anneals.

A recent development consists in gradually lowering the energy of the successive implantation steps in order to fully oxidize the oxygen-rich transition region located above the buried oxide [4]. Multiple SIMOX process is effective in reducing the density of dislocations and precipitates in the film and that of Si islands in the BOX. Recent attention is given to the fabrication of large SIMOX wafers, with ultra-thin Si film and BOX, for ULSI CMOS circuits. Heat dissipation and short-channel effects are improved with thin oxides.

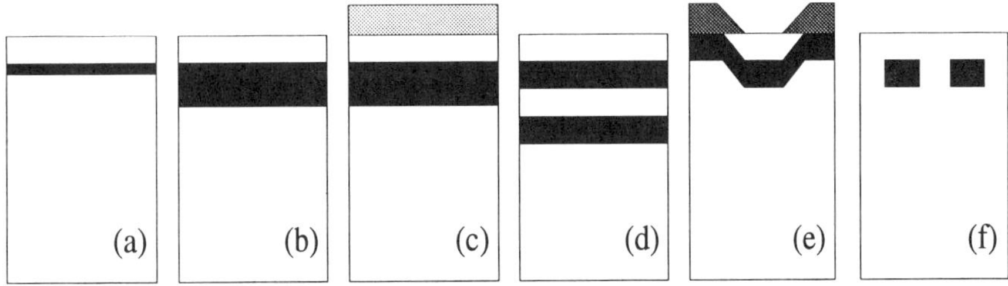

Figure 6.2: Various types of SIMOX structures: (a) thin BOX (lower dose), (b) thin silicon film (lower energy), (c) thick film completed by epitaxy, (d) double SIMOX, (e) isolated single-transistor islands, (f) interrupted BOX.

In regular quality SIMOX, the buried oxide *interfaces* are sharp and uniform. The density of traps and fixed charges at the upper interface of the buried oxide is typically in the range of $D_{it_2} \simeq 0.5$–$2 \times 10^{11}\,\mathrm{cm^{-2}eV^{-1}}$. It is definitely larger than at the interface between film and gate oxide ($D_{it_1} \simeq 10^{10}\,\mathrm{cm^{-2}eV^{-1}}$, as in bulk silicon) but small enough not to jeopardize the performance of integrated circuits. As a matter of fact, the carrier mobility at this interface reaches 80–95% of the front channel mobility. The thin silicon layer is wafer-scale monocrystal with high quality and excellent electrical properties [5]. The interface sharpness, the homogeneity of the buried oxide and the flexibility of engineering the structure represent sound arguments that promote SIMOX as the most suitable technology for thin film applications.

6.3.2 Wafer Bonding

The wafer bonding (WB) technique provides undamaged crystal quality and a wide range of thicknesses for both the SOI film and the buried oxide layer [6]. There are three basic steps which are required for the wafer bonding process [7]: (i) mating two silicon wafers together at room temperature, (ii) annealing the bonded wafers to increase bonding strength, and (iii) thinning one of the two wafers to a proper thickness by grinding and polishing and/or etching. The main drawback of wafer bonding technique is its difficulty in producing extremely thin uniform Si film by the conventional polishing technique. Therefore, the third step is more critical for manufacturing ultrathin Si films with good uniformity. Recently, a plasma assisted chemical etching method (PACE) for fabricating extremely thin (100 nm), uniform SOI films has been developed by Mumola *et al* [8].

The bonding strength is usually very low ($\sim 70 \, \text{erg/cm}^2$) after initial room temperature bonding. In order to increase the bonding strength and maintain the wafers tightly bonded during thinning and device processing, the bonded wafers are annealed at $800\,^\circ\text{C}$ or higher temperature. The bonding strength increases monotonically with temperature and becomes more pronounced above $600\,^\circ\text{C}$. Two procedures are combined to obtain useful bonded SOI substrates for device fabrication. The conventional grinding and *back-polishing* techniques can be applied to achieve SOI layers of $2\,\mu\text{m}$ or thicker, which satisfy the requirement of bipolar devices and power MOS transistors. After polishing down to a few micron thickness, an *etch-back* process is used to prepare thinner SOI layers. The etching thins the SOI film to a desired thickness that is determined by an etch-stop layer [9]. The process makes use of the etching selectivity between the highly doped etch-stop layer and the undoped silicon layer.

6.3.3 Zone Melting Recrystallization

The Zone Melting Recrystallization (ZMR) technology produces SOI structures by recrystallization of polysilicon films deposited on oxidized wafers. In the ZMR process, a thermal oxide (1–$2\,\mu\text{m}$ thick) is first grown on a silicon substrate, followed by deposition of LPCVD amorphous or polycrystalline silicon film (0.5–$1.0\,\mu\text{m}$ thick) on the thermal oxide. The whole structure is capped with a $2\,\mu\text{m}$ thick layer of deposited thermal oxide covered by a thin nitride (Si_3N_4) layer. A melting zone, created by lamps [10], graphite strip heater [11], electron beam [12], or laser is scanned across the entire silicon wafer. As a result, full liquid-phase recrystallization of a silicon wafer can be carried out in a single pass.

The ZMR process can be seeded or unseeded. For seeded ZMR, the oxidized Si wafer is patterned with continuous seeding lines or discrete seeding windows ($1 \times 1\,\mu\text{m}^2$). Seeding is responsible for thermal and topological discontinuities which may induce mass transport. Surface uniformity can be recovered by planarization and thinning. The rapid cooling of the molten zone is responsible for defect generation near the seeds. SOI devices must be located, above the oxide, in "defect-free" islands.

Unseeded ZMR–SOI films are basically composed of large polycrystalline grains. The location of grain boundaries can be controlled by depositing a patterned antireflective Si_3N_4 layer over the silicon film. The grain boundaries are confined underneath the patterns as a result of the discontinuity in thermal gradient. Pseudo-unseeded ZMR is obtained by disposing the seeds on the periphery of the wafer. After the ZMR process is completed, the seeded region is removed and the SOI film looks like a wafer-scale unseeded monocrystal.

The predominant defects that hamper the wide application of ZMR SOI materials are grain subboundaries. The misorientation of the subboundaries and the resulting stress causes the formation of dislocation arrays. The outstanding merit of ZMR process is connected with the production of three-dimensional (3–D) integrated circuits. Potential advantages of 3–D integration include multifunctional operation, facilitation of parallel processing, optical sensing functions, and high-speed.

6.3.4 Full Isolation by Porous Oxidized Silicon

FIPOS technology (Full Isolation by Porous Oxidized Silicon) involves the use of oxidized porous silicon [13, 14]. This technique is more complicated than the SIMOX and ZMR processes, but offers the potential for essentially defect-free silicon layers. Oxidation of the porous silicon layer leads to standard thermal oxide, and does not disturb the high quality of the original silicon film.

In a typical FIPOS sequence, phosphorus islands are implanted in a p-type Si wafer. Anodic reaction in a concentrated hydrofluoric acid (HF) solution is used to convert the superficial p-type regions into porous silicon. The implanted n-type islands are not affected. As porous Si has a very large surface/volume ratio (200–$1000\,\mathrm{m^2}$ per $\mathrm{cm^3}$!), it is selectively oxidized at 750–$800\,^\circ\mathrm{C}$. The final structure is obtained after annealing at $1100\,^\circ\mathrm{C}$ for oxide densification. The disadvantage of FIPOS is that it involves a wet process. The main expectation comes from the possible superposition of two applications of porous silicon: electroluminescence and easy conversion in SOI structures.

6.3.5 Silicon on Sapphire

The epitaxial growth of Si films on $\mathrm{Al_2O_3}$ starts with the formation of small, hemispherical silicon islands whose size is gradually enlarged. When the islands are completely coalesced, the film thickness is about $20\,\mathrm{nm}$ thick. This transition region is characterized by lateral stress (strong silicon-sapphire bonds), many crystallographic defects reflecting the lattice mismatch, and aluminum contamination [15]. The incorporation of Al atoms, released from the substrate, into silicon aggregates explains the autodoping of the transition layer.

More damaging than the lattice mismatch (10%) is the difference in thermal expansion coefficients between Si and $\mathrm{Al_2O_3}$. During the post-growth cooling phase, differential contraction occurs which causes strong compression of the Si film. Only part of the contaminant species and crystallographic defects (dislocations, microtwins, and stacking faults) are eliminated by annealing at about $1100\,^\circ\mathrm{C}$.

The standard SOS material can be improved by *solid-phase epitaxial regrowth* [16]. Silicon ions are implanted to amorphise the film, except for a thin superficial layer. This complete disordering erases the memory of the silicon lattice and damaged interface. Annealing at about 600 °C allows the epitaxial regrowth of the film, starting from the less defective surface region, which plays the role of a seed, towards the Si–Al$_2$O$_3$ interface. This improvement is visible in carrier mobility and lifetime, but also impacts on the cost of already expensive SOS wafers.

The electrical properties of SOS structures are governed by three key aspects: lateral stress, in-depth inhomogeneity of the film, and transition layer. The stress modifies the configuration of silicon energy bands and carrier mobility [17]. The measurements also show that the electrical parameters (relaxation time, mobility, lifetime) are thickness-dependent and degrade in thinner films (see Figure 6.15(b)). The net degradation of the transport parameters near the Si–Al$_2$O$_3$ interface is associated with the presence of the transition layer [18]. Besides the poor crystal quality, the autodoping is of great concern since aluminum contaminant is electrically active.

6.4 Wafer Screening by Ψ-MOSFET Technique

Wafer screening refers to the evaluation of crystal quality, doping uniformity, defect and impurity concentrations, thicknesses of the silicon film and buried oxide, carrier lifetime, mobility, and trapping in starting wafers prior to the fabrication of SOI devices. Fluctuations in quality and uniformity of SOI materials have great impact on the performance, yield, and reliability of the VLSI circuits. Therefore, it is essential that diagnostic tools including both destructive and nondestructive techniques be available for initial assessment of the key physical and electrical parameters of the starting SOI materials. The conventional electrical characterization proceeds from capacitance and conductance measurements and require the fabrication of p-n junctions and capacitor structures, hence it is more suitable for the assessment of processed devices.

Described below is a simple technique which takes advantage of the specific configuration of SOI and has the potential of being non-destructive. The pseudo–MOS transistor, also called point-contact transistor or Ψ–MOSFET, is the first transistor that does not require any lithography or technology at all.

The Ψ–MOSFET is based on the upside-down MOS structure that is inherent in all SOI materials. Figure 6.3 shows that the bulk Si substrate acts as a gate terminal and can be biased, through the metal support, to induce a conduction channel at the upper interface of the buried oxide. The buried oxide plays the role of a gate oxide and the Si film represents the transistor body. To operate *in situ* (without lithography and metallization) the Ψ–MOSFET, low pressure probes are placed on the film and form source and drain point contacts [19, 20].

In spite of the device simplicity and non-parallelism of current lines, very pure MOSFET–like characteristics are produced. As shown in Figure 6.3(b), the output $I_D(V_D)$ curves exhibit a nearly ideal behavior up to 20 V. The locus of I_{Dsat} *vs.* V_{Dsat} (dashed line) indicates that the saturation voltage follows the

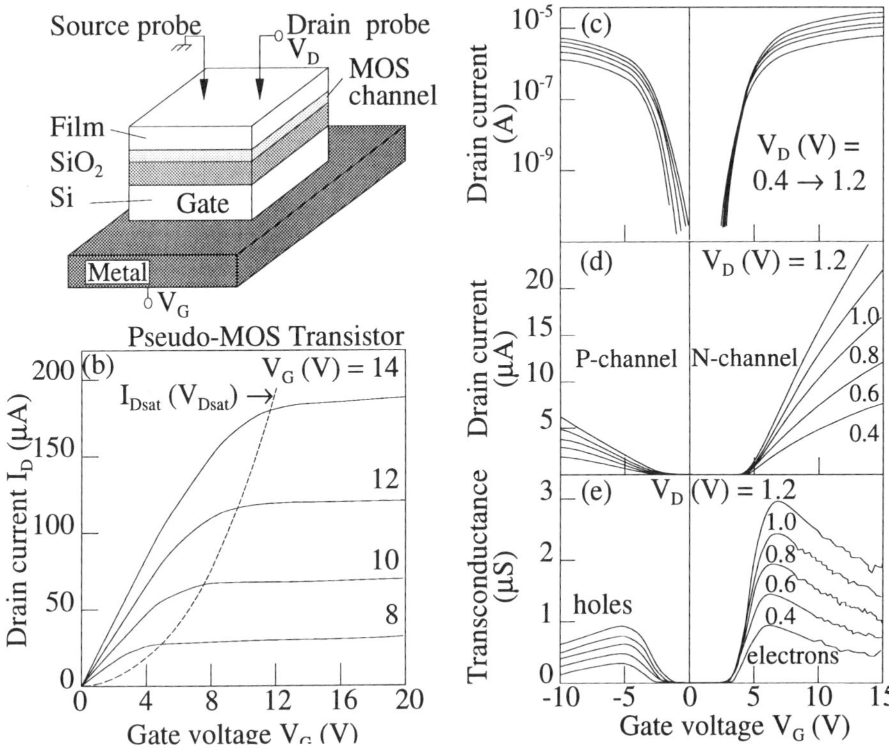

Figure 6.3: Schematic of the pseudo-MOS transistor (a) and typical characteristics: (b) drain current versus drain voltage, (c) subthreshold characteristics, (d) current versus gate voltage in strong inversion and accumulation, (e) transconductance curves.

standard MOS model: $V_{Dsat} \simeq V_G - V_T$.

According to the positive or negative bias applied on the gate, accumulation or inversion channels are activated at the interface. Typical subthreshold $I_D(V_G)$ characteristics are shown for n- and p-channels in Figure 6.3(c). Strong inversion and strong accumulation curves are given in Figure 6.3(d), whereas Figure 6.3(e) illustrates the corresponding transconductance characteristics. From these diagrams, many useful properties may be determined for electrons and holes.

The basic set-up is composed of any two-probe system, connected to a HP–4145 or equivalent pico-ameter. However, for calibration purposes (see Figure 6.5(a)), it is more suitable to use a four-point prober which currently serves for resistivity measurements (for example, the Jandel system has tungsten-carbide probes with pressure control).

Increasing the probe *pressure* gradually reduces the series resistances and

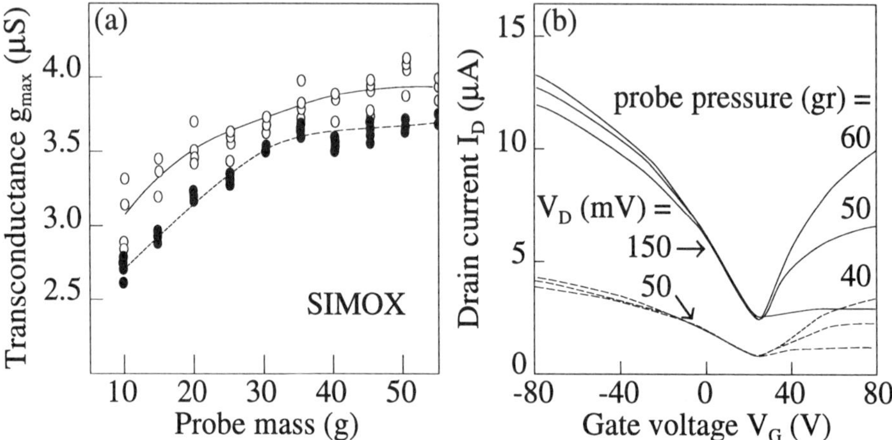

Figure 6.4: Influence of the probe mass (pressure) on (a) the transconductance peak, $g_{max} = f_g \mu_0 C_{ox} V_D$ (- - - from transconductance curves, —— from drain current curves) and on (b) the $I_D(V_G)$ characteristics.

provides more accurate carrier mobility values [19, 21]. In 200 nm thick fully depleted films, it was found that the transconductance improves by 35 % as the pressure increases from 10 g to 30 g, and then saturates (Figure 6.4(a)). Larger pressures are needed in thicker or more heavily doped films which cannot be totally depleted by applying a substrate bias. Figure 6.4(b) shows that the influence of the pressure is quite substantial in inversion ($V_G > 0$), where the tip contact to the n-channel is achieved through a p-doped overlay and a depletion zone. In contrast, the contact to the accumulation p-channel is direct and the pressure effect is much less relevant. Therefore, the pressure adjustment is a matter of trade-off for each type of SOI wafer. The probes must penetrate the native oxide and preferably reach the depletion region, without producing too much damage. The *spacing* between source and drain probes is less critical.

The drain current originates from the inversion/accumulation channel as well as from the neutral region of the film. The accurate modeling of the Ψ–MOSFET requires the solutions of the Poisson and Gauss equations in the film and in the underlying substrate. There are many possible scenarios according to the various combinations of film/substrate doping levels and types, or interface defects.

In the simpler case of *fully depleted* films, I_D is governed by either the inversion or the accumulation channel and the intercepts with V_G–axis roughly correspond to the threshold voltage V_T and flat-band voltage V_{FB}:

$$I_D = f_g C_{ox} \frac{\mu_0}{1 + \theta(V_G - V_{T,FB})}(V_G - V_{T,FB})V_D \tag{6.1}$$

where μ_0 is the electron/hole mobility and θ is the mobility attenuation factor

$$\theta = \theta^0 + f_g \mu_0 C_{ox} R_{SD} \gg \theta^0 \tag{6.2}$$

The ideal value, θ^0, is proportional to C_{ox} and can be very small ($\theta^0 \leq 0.01\,\mathrm{V}^{-1}$) for fully processed back channel SOI MOSFETs. By contrast, in the Ψ–MOSFET, the Schottky nature of the probe contacts results in significant series resistances and larger θ values ($\theta \geq 0.05\,\mathrm{V}^{-1}$).

The geometric coefficient f_g accounts for (i) non-parallel current lines, (ii) lack of lateral isolation, and (iii) size and shape of probe contacts. It is not possible to define the width, length, or aspect ratio of the Ψ–MOSFET, so the usual relation $f_g = W/L$ does not apply and f_g must be independently determined.

In a highly doped p-type film which is *partially depleted*, the total current is the sum of the inversion/accumulation channel current given by Equation 6.1 and the volume current. As the depletion depth w_d shrinks from inversion to accumulation, the volume current increases quasi-linearly with gate voltage

$$I_D = q f_g \mu_0 N_A V_D (t_{si} - w_d) = f_g \mu_0 C_{ox} V_D (V_0 - V_G) \tag{6.3}$$

where

$$V_0 = V_{FB} + \frac{q N_A}{C_{ox}} t_{si} \tag{6.4}$$

V_0 is the fictive voltage that would lead to full depletion. It is obtained from the intercept of the linear current, corresponding to the depletion region with the horizontal axis. It is worth noting that, in this region, the transconductance is constant.

The *geometric coefficient* f_g, which reflects the lateral current spreading, is determined by comparing Ψ–MOSFET data with four-point probe experiments performed with the same system. Applying a current I_{14} between the outer probes, while measuring the voltage drop V_{23} between the inner probes, eliminates the influence of series resistances. This provides well calibrated values for the sheet resistance $R_\square = V_{23}/I_{14}$, which is V_G–dependent too:

$$I_{14} = 4.53 C_{ox} \frac{\mu_0}{1 + \theta(V_G - V_{T,FB})} (V_G - V_{T,FB}) V_{23} \tag{6.5}$$

A direct comparison between Equations (6.1) and (6.5) can be made by plotting both $I_{14}(V_G)$ and $I_D(V_G)$ curves while keeping the voltage V_{23} constant and equal to V_D. The similarity of the two curves is illustrated in Figure 6.5(a). Since I_{14} is 6 times larger than I_D, an empirical value is obtained for the geometrical coefficient $f_g \simeq 0.75$ [19].

The *threshold and flat-band* voltages are determined from $I_D(V_G)$ characteristics. The inversion region is discriminated from the accumulation region by a larger sensitivity to probe pressure. In order to cancel first order series resistance effects, the values of μ_0 and $V_{T,FB}$ are extracted by plotting the function $I_D/\sqrt{g_m}$ versus V_G. The slope, $\sqrt{f_g \mu_0 C_{ox} V_D}$, yields the mobility for electrons

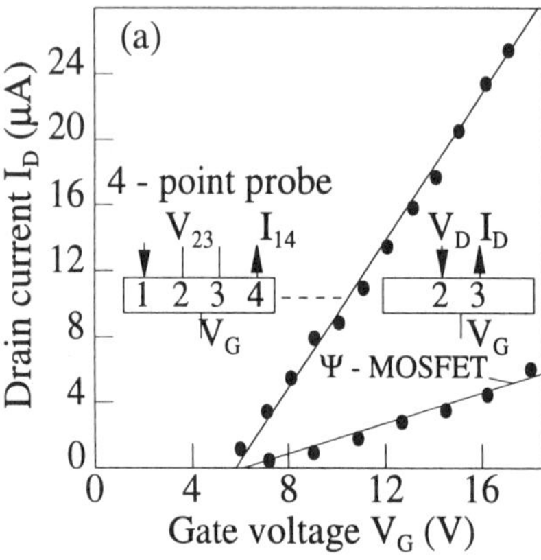

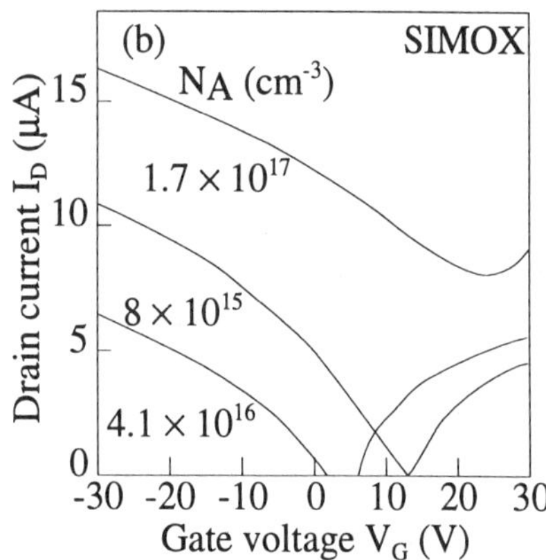

Figure 6.5: (a) Comparison between the variations of the Ψ–MOSFET drain current and four-point probe current versus gate voltage (the voltage drop between the inner probes was $V_D = V_{23} = 0.2\,\text{V}$). (b) Influence of film doping on $I_D(V_G)$ characteristics.

or holes, and the intercept gives V_T or V_{FB}. *Coefficient θ* is readily obtained from Equation 6.1, and the series resistances are determined using Equation 6.2. The voltage V_0 is found by extrapolating to zero the linear portion of the $I_D(V_G)$ characteristic [1].

A simple look taken at the Ψ–MOSFET $I_D(V_G)$ characteristics of Figure 5(b) is enough to detect if the film behaves as fully or partially depleted. *Doping levels* above $5 \times 10^{15} cm^{-3}$ are quite precisely determined by applying Equation 6.4 to the experimental values of V_{FB} and V_0. The characterization of *fixed charges* in the buried oxide proceeds from the analysis of the flat-band voltage. Finally, the density of *interface traps* is determined from the subthreshold swing, $S = (kT/q)(1 + C_d/C_{ox} + qD_{it}/C_{ox})$, or from the difference between V_T and V_{FB}. In summary, the Ψ–MOSFET stands as a very unique method for the rapid characterization of SOI wafers.

6.5 Capacitance and Conductance Techniques

The capacitance and conductance techniques are frequently used for extracting the parameters of Si-SiO$_2$ interface in bulk silicon MOS systems. Although the method becomes heavier, it can still be applied to the silicon-insulator-silicon (SIS) capacitor structure formed on a SOI substrate [22, 23]. The film is biased through a top metal contact (gate) and the back of the wafer is grounded. The conventional MOS capacitor theory is adapted to include the top and bottom Si/SiO$_2$ interfaces of the SOI buried oxide. By introducing a coupling factor, which accounts for the interaction of the two surface potentials, a small signal equivalent circuit can be defined for the SIS capacitor, and then reduced to a MOS-like equivalent circuit.

The total capacitance C_T of an n-i-n SIS capacitor on SOI is expressed by [24, 25]

$$\frac{1}{C_T} = \left(\frac{dQ_T}{dV_G}\right)^{-1} = \frac{1}{C_{s_2} + C_{it_2}} + \frac{1}{C_{ox_2}} + \frac{1}{C_{s_3} + C_{it_3}} \tag{6.6}$$

where $C_s = -dQ_s/d\psi_s$ is the surface capacitance, $C_{it} = -dQ_{it}/d\psi_s$ is the static capacitance of the interface traps, and C_{ox_2} is the buried oxide capacitance. In the following analysis, we consider the case of a negative gate bias which tends to accumulate the upper interface I_2 and deplete the lower interface I_3 of the buried oxide. It is useful to define a coupling factor $K_2(\psi_{s_3})$

$$K_2(\psi_{s3}) \equiv \frac{d\psi_{s2}}{d\psi_{s3}} = -\frac{C_{s3} + C_{it3}}{C_{s2} + C_{it2}} \tag{6.7}$$

which yields

$$\frac{1}{C_T} = \frac{1}{C_{ox_2}} + \frac{1 - K_2}{C_{s_3} + C_{it_3}} \tag{6.8}$$

According to Equation 6.8, the equivalent circuit of a SIS capacitor reduces to a MOS-like equivalent circuit. For $K_2 = 0$, there is no charge coupling and the equivalent circuit for a SIS capacitor is identical to that of a conventional MOS capacitor.

Following the "low-frequency" MOS capacitance method, the interface trap capacitance is derived from

$$C_{it_3}(V_G) = (1 - K_2)\left(\frac{1}{C_{LF}} - \frac{1}{C_{ox_2}}\right)^{-1} - C_{s_3}(V_G) \tag{6.9}$$

When the signal frequency is low enough, then the interface traps do follow the a.c. signal.

The conductance method [26] can also be employed for determining the interface trap density in a SIS capacitor operated in the depletion region. The parallel branch of the equivalent circuit can be converted into a frequency-dependent capacitance C_p and conductance G_p

$$C_p = (1 - K_2)^{-1}\left(C_{s_3} + \frac{C_{it_3}}{1 + \omega^2\tau_{it}^2}\right) \tag{6.10}$$

$$\frac{G_p}{\omega} = (1 - K_2)^{-1}\left(\frac{\omega\tau_{it}C_{it_3}}{1 + \omega^2\tau_{it}^2}\right) \tag{6.11}$$

The above expressions are for a single level interface trap and can be adapted to a continuous distribution of traps throughout the band gap.

Prior to extracting the interface properties, it is necessary to know the SIS capacitor parameters such as buried oxide thickness, doping concentration, and fixed oxide charge. The buried oxide thickness t_{ox_2} can be determined

170

from the measured maximum high-frequency capacitance C_{HF}^{max} [22]. Once the buried oxide thickness is known, the doping concentrations in film and substrate are calculated from the minimum high-frequency capacitance regardless of the interface trap density [1].

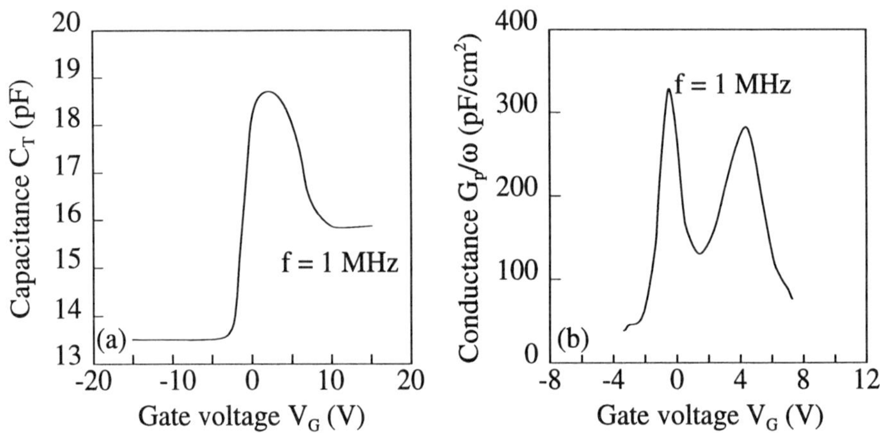

Figure 6.6: High-frequency capacitance and conductance curves for a typical SIMOX n-i-n SIS capacitor.

The fixed oxide charge densities, Q_{f_2} and Q_{f_3}, are determined from the voltage shift of the measured high frequency C-V curve by comparing it with the ideal C-V curve. Applying a gate bias to accumulate one interface yields the fixed oxide charge at the other interface. The stretchout and shift in the C-V curves, observed at positive gate biases in SIMOX capacitors, were attributed to an increase of interface trap density and fixed oxide charge density at the upper oxide interface.

However, this method is less accurate if the thickness of the film is too small such that the film becomes fully depleted before reaching the minimum capacitance. In this case, the substrate doping density N_{D_3} can be deduced from

$$N_{D_3}(w_3) = -2\left\{ q\epsilon_{si}\frac{d}{dV_G}\left[\frac{1}{(1-K_2)C_{HF}^2}\right]\right\}^{-1} \tag{6.12}$$

where w_3 is the distance from the substrate/buried oxide interface. Since the precise knowledge of the value of $(1 - K_2)$ requires the prior evaluation of the doping concentrations in film and substrate, an iteration procedure becomes necessary. A rapid convergence is expected, since $(1 - K_2)$ is relatively insensitive to the doping concentration, especially near the midgap.

6.6 SOI–MOSFETs: Basic Operation and Typical Characteristics

The transistor characteristics represent an invaluable and straightforward source of information on the intrinsic properties of SOI structures as well as on the process-related defects. In this section attention will be paid to the most important phenomena and parameters of SOI transistors. It is clear that the standard expressions existing for bulk Si MOSFETs apply, without any major modification, to SOI transistors provided they have a contact to the Si film (5–terminal devices) and are fabricated in partially depleted films (relatively thick or highly doped). Although the back gate bias acts as an extra experimental parameter, its practical influence on the operation of partially depleted MOSFETs is rather limited. In contrast, totally new relations had to be derived for fully depleted MOSFETs, whose behavior is more complex as it depends on both gate biases.

6.6.1 Interface Coupling

The dual gate configuration of SOI–MOS transistors (Figure 6.1) offers multiple options as how to bias, independently or simultaneously, the two interfaces. In thick partially depleted films, triggering both channels at once results in the total current being just the parallel combination of two separate contributions. The *intrinsic* properties of the front channel are not modified at all by the back gate bias.

The meaning of *interface coupling* is much beyond this simple context of superposed effects. It applies to fully depleted SOI devices, where the pure parameters (threshold voltage, transconductance, interface trap response, etc) of one channel are insidiously affected by the opposite gate. Most experimental results strikingly vary if the opposite interface is biased in accumulation or in inversion. However, the very subtle consequences of interface coupling only appear when at least one of the two interfaces is depleted. For instance, modifying the back gate bias changes *directly* the back surface potential and *indirectly* the apparent properties measured at the front interface; reciprocally, since the front interface potential was altered, this in turn will cause an additional shift of the back channel potential. In this feed-back mechanism, the surface potential is controlled not only by the adjacent gate, but also by the opposite one. Interface coupling also means that front gate measurements are all reminiscent of the applied bias and of the quality of the buried interface and oxide.

Weak interface coupling occurs in films thicker than one depletion region and thinner than the sum of the two depletion zones ($w_{d_{1,2}} < t_{si} < w_{d_1} + w_{d_2}$). More exciting is the case of *strong* coupling, where one gate is enough to deplete the whole film ($t_{si} < w_{d_{1,2}}$). How intense the coupling is, depends on the doping and thickness of the Si overlay.

In bulk Si MOSFETs, the potential bending at the interface results in the carrier confinement to a "triangular" potential well. The quantum energy levels are separated enough for 2–D quantum magneto-transport effects to be observable at very low temperatures. The situation definitely differs in ultra-thin SOI films, where the potential looks as being almost *flat*, if both interfaces are in inversion.

In reality, the film potential is a slow parabolic function of distance. The potential reduction in the middle of the film, as compared to the surface potential, depends linearly on doping and quadratically on film thickness [27]. As a result, the minority carriers are no longer confined to a potential well but to a spatial (geometric) box which actually is the film itself, delimited by the two interfaces. In other words, inversion at the two interfaces automatically initiates *volume inversion* in the whole film [28].

There is good reason to believe that the concept of volume inversion, priviledge of the SOI device family, is an opportunity for improved performance. The increased number of minority carriers allows higher currents to flow, whereas the influence of interface defects and related scattering mechanisms is attenuated offering higher mobilities for "volume" minority carriers. Experimental evidence for volume inversion in SOI MOSFETs has been obtained by biasing the two gates simultaneously [28], or by fabricating special structures: *DELTA* transistor [29] and *gate-all-around* transistor, where a single gate envelops the whole silicon film and biases both interfaces at once [30, 31].

6.6.2 Floating Body Effects

The major parasitic effect in SOI–MOSFETs is related to the build-up of a film charge originated from impact ionization of channel carriers. This charge cannot be removed rapidly enough if no contact is available to the Si film (body). There are various consequences, generically referred to as *floating body effects*.

The *kink effect* denotes an abrupt increase in the saturation current $I_D(V_D)$ of partially-depleted SOI MOSFETs operated in strong inversion (Figure 6.7(a)). In the saturation region, channel carriers generate, by impact ionization near the drain terminal, electron-hole pairs. While the minority carriers are collected by the drain, the majority carriers flow towards the source and give rise to a gradual charge storage in the body. The subsequent increase in the body potential induces the lowering of both the threshold voltage and the source potential barrier. More minority carriers will flow from the source to the channel, thereby causing an excess drain current and producing more pairs through the avalanche process. The signature of this positive feed-back is the sudden increase in current or "kink" in $I_D(V_D)$ characteristics (Figure 6.7(a)).

Fully-depleted MOSFETs are naturally free from kink effect. The disappearance of the kink is explained by the insensitivity of the depletion charge governed by the film thickness to any body potential increase. Unlike the case of partially-depleted MOSFETs, the threshold voltage is constant as long as the amount of stored charge remains reasonable.

In weak inversion, not only is the lateral field stronger for a given V_D, but also the drain current depends exponentially on the body potential. A regenerative physical process is initiated when the body floats, and results in detrimental

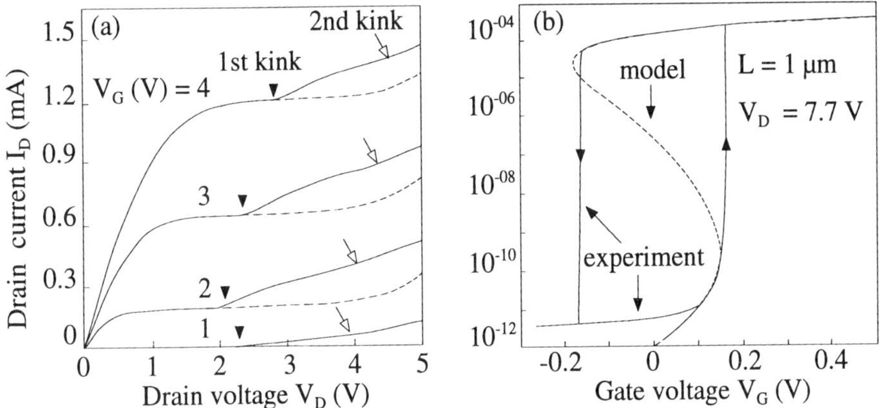

Figure 6.7: Experimental $I_D(V_D)$ and $I_D(V_G)$ characteristics in SOI MOSFETs showing (a) the kink effect (——) which disappears when the body is grounded (- - -) and (b) the hysteresis induced by the negative transconductance region.

consequences. For high drain voltages, the subthreshold characteristics $I_D(V_G)$ become S–shaped. In practice, this *snapback* effect is visualized as a *hysteresis* with anomalously sharp subthreshold slopes (Figures 6.7(b)). Distinct current jumps are recorded for the two directions of the voltage scan, their separation being controlled by the drain voltage magnitude[1].

For a decreasing V_G sweep at high V_D (above a *holding* voltage), the transistor may fail to turn off even for $V_G \leq 0$. Such a complete loss of gate control is called *single transistor latch*. In fact, S–shaped $I_D(V_D)$ characteristics can be experimentally obtained by imposing the current and V_G, and measuring the corresponding drain voltage. A reasonable analytical model was constructed by considering simple relations for the physical mechanisms involved in the positive feed-back loop [32].

In short channel MOSFETs, the floating body effects are amplified by the activation of the *lateral bipolar transistor*. When the source to body junction is on, the bipolar transistor comes into play by injecting additional minority carriers from the source (emitter) into the body (base). The gain of the bipolar transistor, β, depends on channel length and diffusion length, whereas the multiplication factor M is an exponential function of drain voltage. Breakdown occurs when the drain voltage becomes high enough to satisfy the condition

$$\beta \times (M - 1) = 1 \qquad (6.13)$$

This relation indicates that the bipolar transistor action is responsible for the

---

[1] It is interesting to note that negative values of the conductance are obtained simultaneously with those of the negative transconductance.

174

premature breakdown in SOI as compared to bulk Si transistors. The feedback mechanism is self-limiting because an increased body potential leads to a higher saturation voltage which in turn lowers the peak of the lateral field and $(M-1)$. The breakdown voltage degrades with decreasing V_G, which is opposite to the behavior of bulk Si or thick SOI MOSFETs [33, 34]. This difference is explained by accounting for the source junction turn-on and avalanche multiplication.

6.6.3 Transients Effects

The total isolation of the transistor body is often responsible for long current transients. Once the body is charged by impact ionization or gate voltage pulses, a long relaxation time is needed to recombine the stored charge and reach equilibrium. Conversely, if a prompt supply of holes is demanded in an n-channel MOSFET, there is no solution other than the generation mechanism to create them. The amplitude of such transient effects is, therefore, governed by the carrier lifetime in the material.

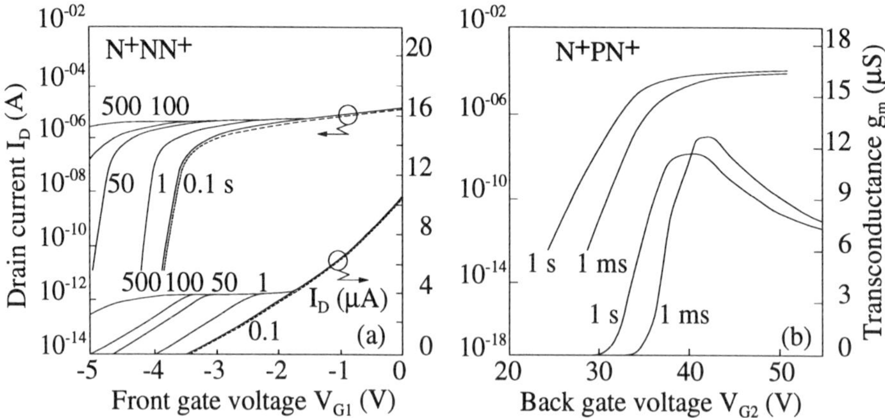

Figure 6.8: $I_D(V_G)$ characteristics showing the influence of (a) the hold time in accumulation-mode MOSFETs (only the dotted curve was scanned from positive to negative bias) and (b) the delay time in enhancement-mode partially-depleted MOSFETs.

The so-called "static" $I_D(V_G)$ characteristics are in general measured with automated systems, so that the data acquisition is much faster than the generation-recombination time constants in good quality SOI films. The $I_D(V_G)$ curves can be completely misunderstood if the transients effects are overlooked. In a partially depleted *depletion-mode* N$^+$NN$^+$ MOSFET, the *ohmic* $I_D(V_G)$ characteristics may depend on the direction of voltage scanning. If V_G is varied from accumulation to inversion (negative sweep—dotted curve in Figure 6.8(a)), small *deep-depletion* regions are formed at each bias step. The temporary absence of steady-state depletion regions renders the current drop exaggerated and inaccurate curves are recorded, unless long delay times are allowed between the measured points.

Therefore, a positive V_G sweep is normally used, but even so there is no certitude of accuracy. Figure 6.8(a) shows that the characteristics recorded *immediately* after a starting negative voltage was applied to the gate are typical for a totally depleted MOSFET. This is simply because a large deep depletion region was initially formed and temporarily covers the whole film. Indeed, there is no reservoir of holes to annihilate promptly the deep-depletion region. In contrast, if once the bias was applied a hold time as long as 500 sec is allowed before the curve is recorded, the minority carrier generation does succeed in restoring equilibrium: the transistor turns out to be partially depleted ... The confusion between a partially- and a fully-depleted transistor is misleading. For instance, the doping of the film is straightforwardly underestimated. Subsequently, using the right resistivity value, yields of course an overestimated mobility.

There are symmetrical situations in enhancement-mode transistors, due to the absence of majority carrier reservoirs. The "static" characteristics of n–MOSFETs may be affected by the direction of voltage scanning, hold time and delay time, as soon as a supply or removal of holes is necessary for reaching equilibrium. Figure 6.8(b) shows that the subthreshold slope, threshold voltage, and transconductance depend on the experimentalist patience.

A drain current *overshoot* is observed in partially-depleted SOI transistors when the gate is switched from accumulation to inversion. Majority carriers, expelled from the surface, temporarily increase the body potential and lower the threshold voltage. The turn-on current may exceed twice the steady-state value. The overshoot amplitude ΔI_D is proportional to the number of excess holes and therefore increases for a higher doping or for a stronger accumulation before turn-on. Monitoring the current overshoot allows determining the recombination lifetime. On the other hand, it is a source of parasitics in DLTS measurements and circuit operation.

The thermal conductivity of SiO_2 layers that inherently surround any SOI device is about two orders of magnitude poorer than for Si. This inhibits the cooling of SOI transistors and causes self-heating problems. A *negative differential conductance*, $\partial I_D/\partial V_D < 0$, is observed for high gate voltages. As V_D increases in the saturation region, the channel temperature rises and makes the carrier mobility drop ($\mu \sim T^{-1.5}$). The rise of the silicon island temperature can be visualized by thermal imaging [35] or by monitoring the subsequent variations of junction leakage current and poly-Si gate resistance.

The channel temperature is proportional to the power dissipated in the device. Temperature raises exceeding $100\,^{\circ}C$ were observed in submicron transistors under static conditions of operation [36]. The reduction in the saturation current and the negative conductance value depend on the direction of the drain voltage and sweep rate. Systematic measurements show that self-heating is more severe for thinner films, shorter channels and thicker buried oxides.

176

6.6.4 Threshold Voltage

The model developed by Lim and Fossum [37] for the case of *fully depleted* enhancement-mode n-channel MOSFETs is based on several hypotheses: long channel, constant doping, charge sheet approximation, no influence from the third interface located below the buried oxide. Integration of the Poisson equation over the Si film yields

$$E_{s_1} - E_{s_2} = \frac{qN_A t_{si}}{\epsilon_{si}}; \qquad \psi_{s_1} - \psi_{s_2} = t_{si} E_{s_1} - \frac{qN_A t_{si}^2}{2\epsilon_{si}} \qquad (6.14)$$

where $E_{s_{1,2}}$ and $\psi_{s_{1,2}}$ stand for the vertical fields and potentials (defined by assuming a virtual neutral point in the film) at the front and back interfaces. The relations between gate voltages and surface potentials are

$$V_{G_{1,2}} = \psi_{s_{1,2}} + \psi_{ox_{1,2}} + \phi_{ms_{1,2}} \qquad (6.15)$$

where $\phi_{ms_{1,2}}$ are metal-semiconductor work functions. The potential drops in the oxides $\psi_{ox_{1,2}}$ are obtained by applying the Gauss theorem at both interfaces

$$\psi_{ox_1} = \frac{1}{C_{ox_1}} \left(\epsilon_{si} E_{s_1} - Q_{f_1} - Q_{c_1} + qD_{it_1} \psi_{s_1} \right) \qquad (6.16)$$

$$\psi_{ox_2} = -\frac{1}{C_{ox_2}} \left(\epsilon_{si} E_{s_1} - qN_A t_{si} + Q_{f_2} + Q_{c_2} - qD_{it_2} \psi_{s_2} \right) \qquad (6.17)$$

with $Q_{f_{1,2}}$ being fixed charges in the oxides, $Q_{c_{1,2}}$ surface channel charges and $D_{it_{1,2}}$ average concentrations of interface traps. Eliminating $\psi_{ox_{1,2}}$ and $E_{s_{1,2}}$ between Equations (6.14–6.17), yields

$$V_{G_{1,2}} = \phi_{fb_{1,2}} + \psi_{s_{1,2}} \left(1 + \frac{C_{si} + C_{it_{1,2}}}{C_{ox_{1,2}}} \right) - \psi_{s_{2,1}} \frac{C_{si}}{C_{ox_{1,2}}} - \frac{2Q_{c_{1,2}} + Q_{si}}{2C_{ox_{1,2}}} \qquad (6.18)$$

where $Q_{si} = -qN_A t_{si}$ is the depletion charge, $C_{si} = \epsilon_{si}/t_{si}$ is the film capacitance, and $C_{it_{1,2}} = qD_{it_{1,2}}$ are interface trap capacitances.

The conventional definition of the front channel threshold voltage V_{T_1} implies that $\psi_{s_1} = 2\phi_F$ and $Q_{inv_1} = 0$ in Equation 6.18. There are three generic cases according to the back gate bias:

(1) Back interface accumulated ($\psi_{s_2} = 0$):

$$V_{T_1}^{acc} = \phi_{fb_1} + \left(1 + \frac{C_{si} + C_{it_1}}{C_{ox_1}} \right) 2\phi_F - \frac{Q_{si}}{2C_{ox_1}} \qquad (6.19)$$

(2) Back interface inverted ($\psi_{s_2} = 2\phi_F$):

$$V_{T_1}^{inv} = \phi_{fb_1} + \left(1 + \frac{C_{it_1}}{C_{ox_1}} \right) 2\phi_F - \frac{Q_{si}}{2C_{ox_1}} \qquad (6.20)$$

(3) Back interface depleted ($0 < \psi_{s_2} < 2\phi_F$). Eliminating ψ_{s_2} between Equations (6.18) yields

$$V_{T_1}^{dep} = V_{T_1}^{acc} - \frac{C_{si}C_{ox_2}}{C_{ox_1}(C_{ox_2} + C_{si} + C_{it_2})}\left(V_{G_2} - V_{G_2}^{acc}\right) \qquad (6.21)$$

$V_{G_2}^{acc}$ represents the limiting voltage below which the back interface is always accumulated whatever the values of V_{G_1}

$$V_{G_2}^{acc} = \phi_{fb_2} - \frac{C_{si}}{C_{ox_2}}2\phi_F - \frac{Q_{si}}{2C_{ox_2}} \qquad (6.22)$$

Similarly, the back interface is always inverted for $V_{G_2} > V_{G_2}^{inv}$; the boundary voltage $V_{G_2}^{inv}$ is nothing but the back channel threshold voltage $V_{T_2}^{inv}$

$$V_{G_2}^{inv} = V_{T_2}^{inv} = \phi_{fb_2} + \left(1 + \frac{C_{it_2}}{C_{ox_2}}\right)2\phi_F - \frac{Q_{si}}{2C_{ox_2}} \qquad (6.23)$$

The meaning of the various expressions derived above is illustrated in Figure 6.9. The front channel threshold voltage decreases linearly between two plateaus corresponding to accumulation and inversion at the back interface. The difference between the plateaus, $\Delta V_{T_1} = (C_{si}/C_{ox_1})2\phi_F$, slightly depends on doping, whereas the slope of $V_{T_1}^{dep}(V_{G_2})$ curves does not. These parameters do not reflect the front interface quality, but can be controlled by adjusting the thicknesses of the SOI structure. In ultra thin films, the slope is merely given by the ratio t_{ox_1}/t_{ox_2}. The threshold voltage characteristics, $V_{T_1}(V_{G_2})$, are therefore useful to determine the film and buried oxide thicknesses as well as the density of back interface traps.

We must insist on the polyvalence of Equations (6.19–6.21) as compared to the simpler case of partially depleted films, where the threshold voltage is insensitive to V_{G_2} bias and takes the same form as in bulk Si MOSFETs

$$V_{T_1} = \phi_{fb_1} + \left(1 + \frac{C_{it_1}}{C_{ox_1}}\right)2\phi_F + \frac{\sqrt{4q\epsilon_{si}N_A\phi_F}}{C_{ox_1}} \qquad (6.24)$$

It is clear that symmetrical expressions, obtained by interchanging the subscripts 1 and 2, hold for the back channel threshold voltage. The extension to p-channel or accumulation-mode SOI–MOSFETs is also straightforward [1].

Figure 6.10(a) shows typical "dual-V_T" curves, where the front and back threshold voltages are plotted against opposite gate biases. In agreement with the above model, it is noted that thinner the film, wider the V_T variation range [34]. The experiment confirms that fully depleted transistors are less vulnerable to short-channel effects than partially depleted SOI or bulk silicon MOSFETs. According to Figure 6.10(b), the threshold voltage roll-off in deep-submicron SOI MOSFETs is remarkably attenuated when using ultra-thin silicon films and buried oxides [38].

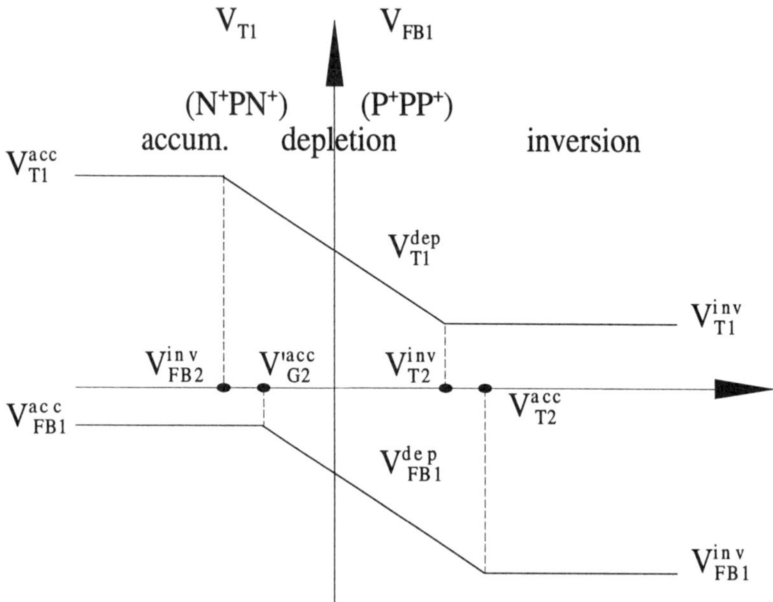

Figure 6.9: Schematic influence of back gate bias on the front channel threshold voltage of enhancement-mode n-channel MOSFETs (N$^+$PN$^+$) and flat-band voltage of accumulation-mode p-channel MOSFETs (P$^+$PP$^+$) in fully depleted SOI MOSFETs.

6.6.5 Subthreshold Slope

In weak inversion, the drain current depends exponentially on gate voltage. The subthreshold slope indicates the sharpness of the current transition from off-state to on-state, and the subthreshold *swing* S is defined as the inverse of the slope. The standard relation for bulk silicon also holds in partially depleted SOI films

$$S = \left(\frac{d \log_{10} I_D}{dV_G} \right)^{-1} = 2.3 \frac{kT}{q} \frac{dV_G}{d\psi_s} = 2.3 \frac{kT}{q} \frac{C_{ox} + C_d + C_{it}}{C_{ox}} \tag{6.25}$$

In fully depleted films, the current depends on both gate biases and can flow at either interface. The inversion charge in weak inversion, the potential fluctuations and the short-channel effects are neglected. For depletion at the back interface, the front channel subthreshold swing is obtained from Equations (6.14–6.18)

$$S_1^{dep} = 2.3 \frac{kT}{q} \left(1 + \frac{C_{it_1}}{C_{ox_1}} + \alpha_1 \frac{C_{si}}{C_{ox_1}} \right) \tag{6.26}$$

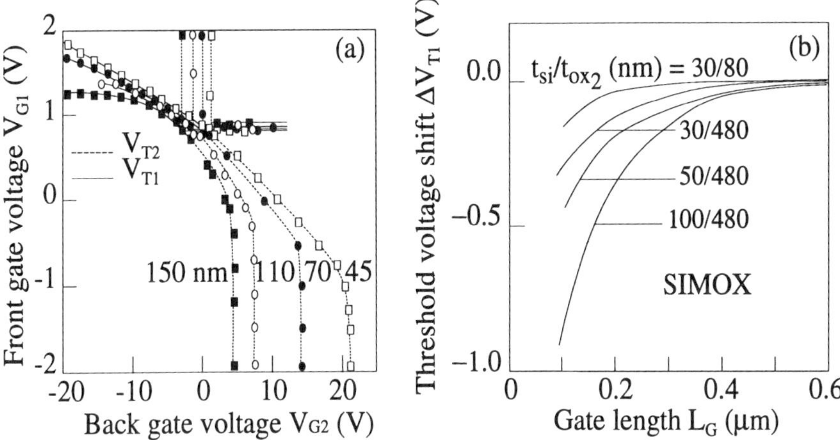

Figure 6.10: (a) Experimental variations of the front channel and back channel threshold voltages versus opposite gate bias in accumulation-mode SIMOX MOSFETs with various thicknesses (after Faynot et al). (b) Threshold voltage reduction against channel length in SOI transistors with more or less thin film and buried oxide (after Ohmura et al).

where α_1 is an interface coupling coefficient

$$\alpha_1 = \frac{C_{ox_2} + C_{it_2}}{C_{si} + C_{ox_2} + C_{it_2}} < 1 \tag{6.27}$$

which accounts for the influence of back interface traps and buried oxide thickness on the front channel current [39, 40].

According to Equation 6.27, thicker buried oxides are beneficial: α_1 is reduced and the swing is improved. In the ideal case, where $D_{it_{1,2}} \simeq 0$ and the buried oxide is much thicker than both the film and the gate oxide, S_1^{dep} approaches the theoretical limit

$$S_1^{min} = 2.3 \frac{kT}{q} \left(1 + \frac{t_{ox_1}}{t_{ox_2}} \right) \simeq 2.3 \frac{kT}{q} \tag{6.28}$$

which corresponds, at room temperature, to 60 mV per decade of current. Accumulation at the back interface decouples the front inversion channel from back interface defects. Indeed, the rapidly increasing capacitance of the accumulation layer makes α_1 tend to unity, degrading the swing.

Front and back gate subthreshold characteristics are shown in Figure 6.11 for an ultra-thin SOI MOSFET. The steeper slopes are always obtained with the opposite interface in depletion. The excellent front channel swing ($S_1^{dep} = 69$ mV/decade) demonstrates the high quality of the two interfaces.

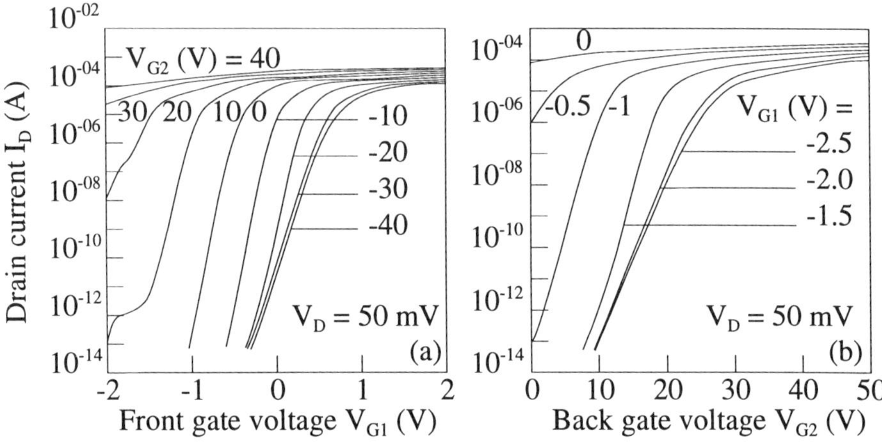

Figure 6.11: Subthreshold current measured as a function of the front (a) or back (b) gate voltage for various biases on the opposite gate (wafer-bonded SOI, $t_{si} = 70\,\mathrm{nm}$, $t_{ox_1} = 27\,\mathrm{nm}$, $t_{ox_2} = 850\,\mathrm{nm}$, $V_D = 50\,\mathrm{mV}$, after Mazhari et al).

6.6.6 Transconductance

Straightforward adaptation of bulk Si relations yields the drain current in a front channel SOI MOSFET, operated in strong inversion and ohmic region

$$I_D = \frac{C_{ox_1} W V_D}{L} \mu_{eff_1} \left(V_{G_1} - V_{T_1}(V_{G_2}) \right) \tag{6.29}$$

where μ_{eff} is the *effective* mobility, function of the vertical field, surface charge and interface defects. The transconductance g_m is given by

$$g_{m_1} = \frac{C_{ox_1} W V_D}{L} \mu_{fe_1} = \frac{C_{ox_1} W V_D}{L} \times \frac{\mu_1}{\left[1 + \theta_1 (V_{G_1} - V_{T_1}(V_{G_2})) \right]^2} \tag{6.30}$$

Here, μ_{fe} is the field effect mobility, μ_1 is the pure mobility of front channel carriers, and θ is the mobility attenuation coefficient in strong inversion.

The complexity of the transconductance in fully depleted MOSFETs (Figure 6.12(a)) is due to the influence of the back gate. Not only does it govern the front channel threshold voltage $V_{T_1}(V_{G_2})$, but also it allows a back inversion channel to be activated. Following the model proposed by Ouisse *et al* [41], there are 4 typical situations as V_{G_2} is varied from negative to positive values.

Case 1 : Back Interface Accumulated

The back surface potential is nearly constant and the threshold voltage and vertical field saturate. This explains why an unique transconductance curve is obtained below a certain bias ($V_{G_2} < -4\,\mathrm{V}$, in Figure 6.12(a)).

Case 2 : Back Interface Depleted

Since V_{T_1} decreases linearly with increasing V_{G_2} (see Equation 6.21), the position of the transconductance peak, $V_{G_1}^{max}$, is laterally shifted. The experiment verifies that $V_{G_1}^{max}(V_{G_2})$ follows the variation of $V_{T_1}(V_{G_2})$, being slightly larger by 0.2–0.3 V. Moreover, the electric field at the front interface decreases, thus allowing the carrier mobility μ_1 and the transconductance peak g_{max_1} to increase.

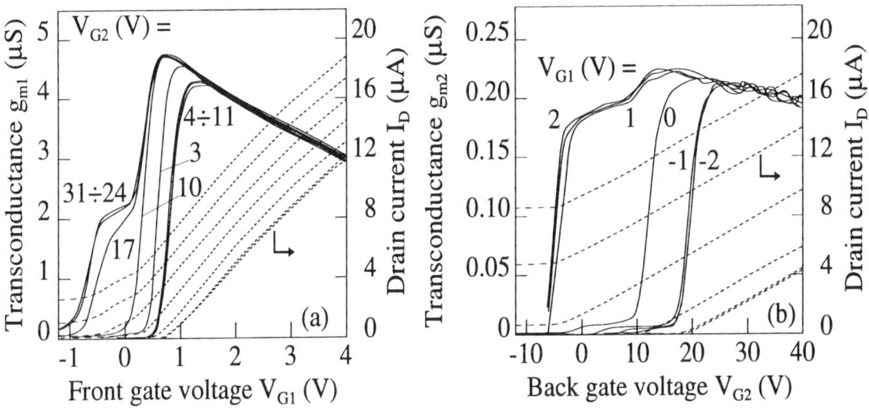

Figure 6.12: Drain current (- - -) and transconductance (—) curves versus (a) front gate voltage and (b) back gate voltage in a long SOI MOSFET ($L = 20\,\mu$m, $V_D = 50\,$mV, $t_{ox_1} = 17\,$nm, $t_{ox_2} = 380\,$nm, $t_{si} = 80\,$nm, $N_A \simeq 5 \times 10^{16}\,$cm^{-3}).

Case 3 : Back Interface from Depletion to Inversion: $V_{T_2}^{inv} < V_{G_2} < V_{T_2}^{acc}$

For a negative V_{G_1} bias, the front interface is accumulated, the back interface is depleted and the drain current is zero. Increasing V_{G_1}, first opens the *back* channel. This happens because, for $V_{G_1} > V_{G_1}^{acc}$, the back threshold voltage decreases and eventually crosses the fixed value V_{G_2} (Figure 6.13(a)). During the back channel conduction, the current is given by

$$I_D = I_{D_2} = \frac{C_{ox_2} W V_D}{L} \frac{\mu_2}{1 + \theta_2 (V_{G_2} - V_{T_2})} (V_{G_2} - V_{T_2}(V_{G_1})) \tag{6.31}$$

where the coefficient $\theta_2 \simeq 5 \times 10^{-3}\,$V^{-1} is extremely small owing to the large thickness of the buried oxide. Remark that although V_{G_2} is fixed, the back current linearly varies with V_{G_1} (Figure 6.13(b)), via the decrease of $V_{T_2}(V_{G_1})$, as long as the front interface remains depleted. Using the reciprocal of Equation 6.21, I_{D_2} can be explicitly rewritten as a function of V_{G_1}

$$I_{D_2} = \frac{\gamma_2 C_{ox_2} W V_D}{L} \frac{\mu_2}{1 + \gamma_2 \theta_2 (V_{G_1} - V'_{T_1})} (V_{G_1} - V'_{T_1}) \tag{6.32}$$

182

where V'_{T_1} stands for a pseudo-threshold voltage

$$V'_{T_1} = V^{acc}_{G_1} + \frac{1}{\gamma_2}(V^{acc}_{T_2} - V_{G_2}) \qquad (6.33)$$

and γ_2 is a threshold voltage coupling coefficient

$$\gamma_2 = \frac{C_{ox_1} C_{si}}{C_{ox_2}(C_{ox_1} + C_{si} + C_{it_1})} \qquad (6.34)$$

The transconductance now expresses the variation of the back channel current under control of V_{G_1}

$$g_{m_1} = \frac{\gamma_2 C_{ox_2} W V_D}{L} \times \frac{\mu_2}{\left[1 + \gamma_2 \theta_2 (V_{G_1} - V'_{T_1})\right]^2} \qquad (6.35)$$

The mobility attenuation factor $\gamma_2 \theta_2$ is very small, hence g_{m_1} is represented by a "plateau" (Figures 6.12(a) and 6.13(c))

$$g_{plat_1} = \frac{\gamma_2 \mu_2 C_{ox_2} W V_D}{L} \qquad (6.36)$$

The plateau of the front channel transconductance is a direct measure of the back channel mobility.

For $V_{G_1} \geq V^{inv}_{G_1}$, the front channel is activated as well, whereas the values of V_{T_2} and I_{D_2} saturate. The total current, now becomes a superposition of I_{D_1} and I_{D_2}, clearly reflected by a slope change in $I_D(V_{G_1})$ characteristics (Figures 6.12(a) and 6.13(b)). The transconductance increases sharply with V_{G_1}, being fully representative for the front channel mobility. This interesting exchange of dominant roles between the front and back interfaces is a unique feature of fully-depleted SOI transistors. The most interesting consequence is the gradual deformation of transconductance curves (Figure 6.12(a)), which become totally different from the conventional curve in bulk silicon.

Case 4 : Back Interface Inverted $V_{G_2} > V^{acc}_{T_2}$

The back interface is always inverted, whatever the bias V_{G_1}. When the front interface is accumulated, the current I_{D_2} is a constant (Figure 6.13(b')) and the transconductance is zero. As soon as $V_{G_1} \geq V^{acc}_{G_1}$, the back channel threshold voltage begins decreasing and I_{D_2} increases linearly with V_{G_1} according to Equation 6.32. The transconductance plateau of Figure 6.13(c') is again given by Equation 6.36, until the front interface reaches strong inversion (for $V_{G_1} > V^{inv}_{G_1}$) and recovers the current control.

Cases 3 and 4 look similar, except that in the latter situation the current is never zero and the extension of the transconductance plateau region is maximum: $V^{acc}_{G_1} \leq V_{G_1} \leq V^{inv}_{G_1}$. From the transconductance curve of Figure 6.13(c'), many useful parameters can directly be deduced: voltages $V^{inv}_{G_1}$, $V^{acc}_{G_1}$, $V^{max}_{G_1} \simeq V_{T_1}$, front and back channel mobilities[2].

---

[2] The series resistances induce a reduction of the effective voltage drop on the drain and are responsible for the apparent degradation of the mobility and coefficient θ.

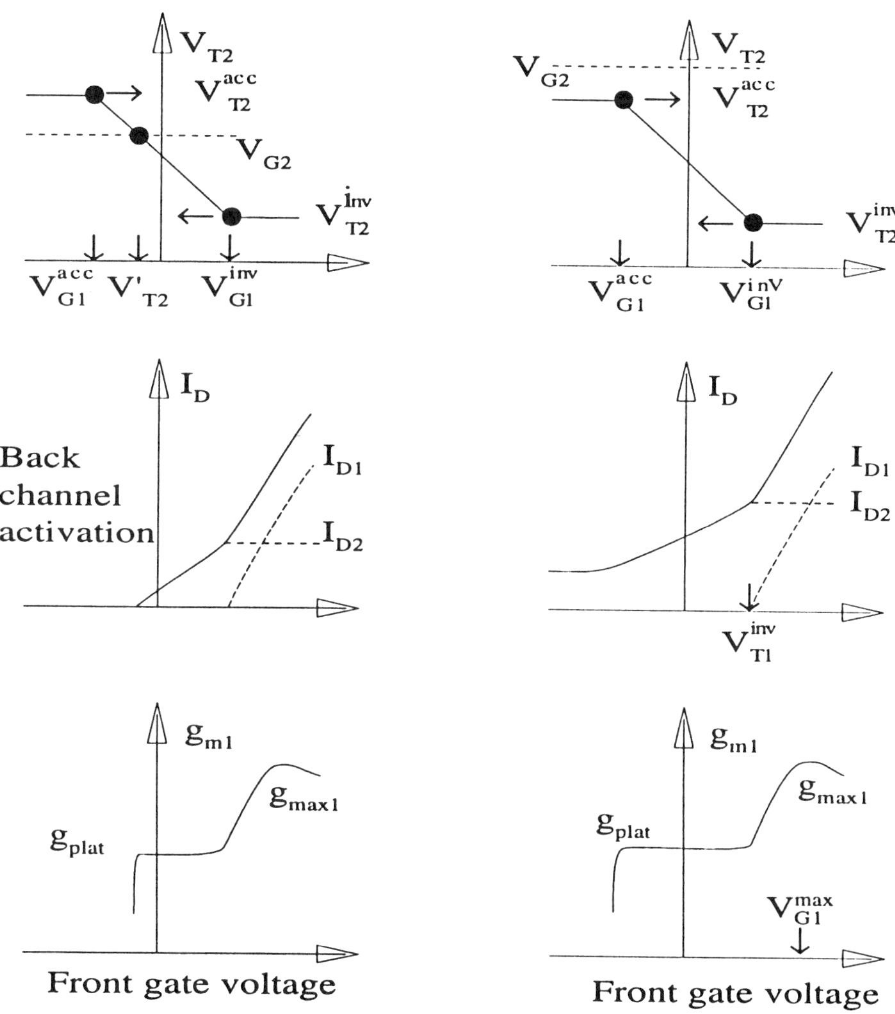

Figure 6.13: Representation of the back channel threshold voltage (a,a'), drain current (b,b') and transconductance (c,c') variations versus front gate bias. Diagrams (a,b,c) correspond to *Case 3* (back channel triggered for $V_{G_1} \geq V'_{T_1}$), whereas (a',b',c') relate to the *Case 4* (back channel always activated) (after Ouisse et al).

The parameter extraction proceeds as in bulk Si MOSFETs, for instance by drawing $I_D/\sqrt{g_m}$ vs. V_G curves. The intercept gives the threshold voltage and the slope yields the carrier mobility in the channel. Another technique consists in taking the second derivative of $I_D(V_G)$ curves. The peaks correspond to the threshold voltage position [1].

6.7 Profiling the Vertical Inhomogeneities

The properties of thin SOI films may vary with distance from the top surface to the buried insulator. Such a *vertical inhomogeneity* arises from defect generation that takes place near the film–insulator interface during the material synthesis (epitaxial growth, oxygen/nitrogen implantation, annealing, etc) [17].

In non-uniform films, the average values of resistivity, mobility and residual doping are meaningless, since they may look very degraded, even if the properties of the top portion of the layer reaches bulk Si standards. Contrasting and misleading average parameters may be obtained in SOI layers with identical quality but different thickness. A very efficient profiling method combines the MOS field effect with magnetoelectric measurements.

Figure 6.14(a) shows the configuration of an MOS-Hall device, which is a 7-terminal depletion-mode n-channel MOSFET. Besides the gate and the two end contacts (source and drain), there are 4 lateral N^+ contacts. The longitudinal voltage drop, measured between contacts $H_{1,2}$ (or $H_{3,4}$), gives the conductance and magnetoresistance values, free of the influence of source/drain junctions. Contacts $H_{2,4}$ (or $H_{1,3}$) serve for Hall voltage measurements.

Hall effect and magneto-conductance experiments are performed at low drain bias as a function of V_G. By gradually depleting the film, the thickness w of the conducting region is reduced and the measured (average) transport properties are modified. A procedure involving differentiation provides the contribution of the infinitesimal layer Δw, situated at the limit between depletion and "active" regions.

The gate voltage dependence of the depletion depth w_d is determined from small-signal high-frequency $C(V_G)$ capacitance measurements

$$w_d = \epsilon_{si} l_x l_y \left(\frac{1}{C(V_G)} - \frac{1}{C_{ox}} \right) \tag{6.37}$$

The vertical profile of a parameter A consists of successive local values $A^*(z)$ which are extracted from the experimental average values $\overline{A}(V_G)$. For example, the average resistance (or magneto-resistance) of a conductive sheet of thickness w is

$$\overline{R}(V_G) = \frac{l_x}{l_y} \left(\int_0^w \sigma^*(z)dz \right)^{-1} \tag{6.38}$$

Conversely, the local conductivity at $z = w$ is given by

$$\sigma^*(z = w) = \frac{l_x}{l_y} \frac{d\overline{R}^{-1}}{dV_G} \frac{dV_G}{dw} \tag{6.39}$$

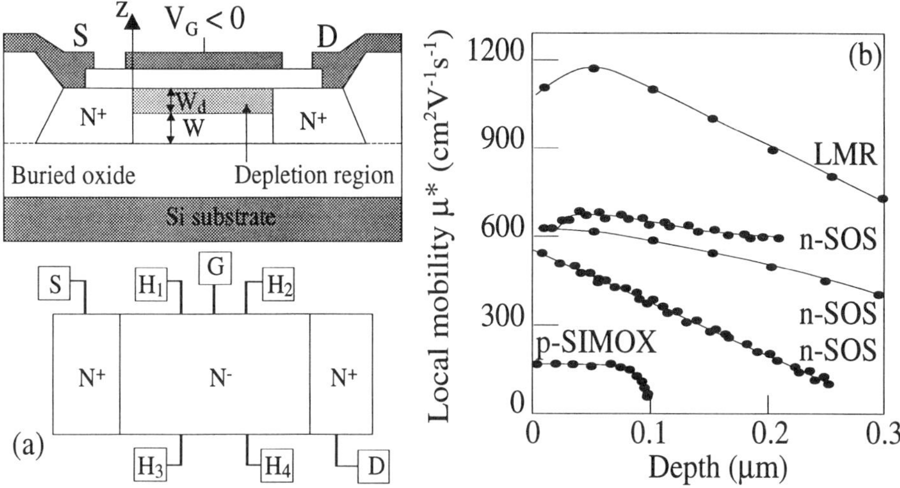

Figure 6.14: (a) Configuration of a seven-terminal MOS-Hall device and (b) carrier mobility profiles (after Lee et al).

The measured average Hall mobility is expressed as [42, 43]

$$\overline{\mu}_H(V_G) = \left(\int_0^w \mu_H^*(z)\sigma_0^*(z)dz \right) \left(\int_0^w \sigma_0^*(z)dz \right)^{-1} \qquad (6.40)$$

where $\sigma_0^*(z)$ is the conductivity profile at zero magnetic field. Differentiating Equation 6.40 and using Equation 6.39 yields the local Hall mobility

$$\mu_H^*(w) = \frac{l_x}{l_y} \frac{1}{\sigma_0^*} \frac{d\left(\overline{\mu}_H \overline{R}_0^{-1}\right)}{dV_G} \frac{dV_G}{dw} = \overline{\mu}_H - \overline{R}_0(V_G)\frac{d\overline{\mu}_H}{dR_0} \qquad (6.41)$$

Most experiments have been conducted on early SOS and SIMOX films, where the material homogeneity used to be a critical technology issue. Typical profiles, derived with Equation 6.39, show in Figure 6.14(b) the mobility degradation in SOS layers. The mobility decreases almost linearly with depth in SOS, whereas in low temperature annealed SIMOX it is flat in the upper half of the film and then degrades rapidly close to the buried interface. In recent SIMOX material, the transport coefficients are quasi-homogeneous. This is simply confirmed by transistor measurements which yield comparable values for the carrier mobility at the front and back channels.

6.8 Charge Pumping Technique

Charge pumping (CP) stands as a very sensitive method for the characterization of low concentrations of interface traps (10^9 cm^{-2}eV^{-1}) in short-channel MOS devices [44]. The adaptation of CP to SOI transistors requires a contact to the Si film [45, 46]. Either 5–terminal MOSFETs or gate-controlled p-i-n diodes may be used [47]. Figure 6.15(a) shows that the two terminals of the $P^+P^-N^+$ diode have different functions. The N^+ contact controls the charge of the inversion layer and can be more or less reversed biased. The grounded P^+ terminal supplies the majority carriers and therefore plays the same role as the body contact in 5–terminal transistors. The experimental procedure and the CP model are described in another chapter of this book. We will focus on the specifics of the technique in SOI.

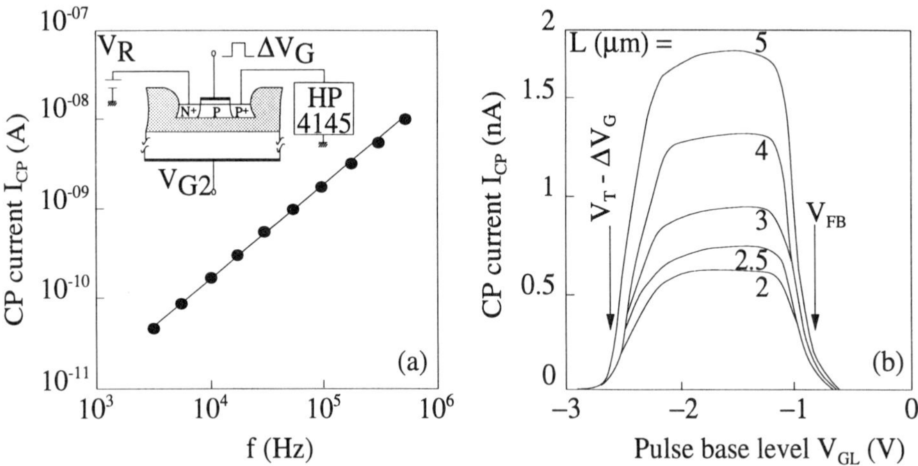

Figure 6.15: (a) Configuration of CP measurements in SOI diodes and variation of CP current versus frequency ($\Delta V_G = 5\,\text{V}$, $V_{GL} = -2\,\text{V}$, $V_R = 0$, $T_R = T_F = 0.1\,\mu\text{s}$); (b) CP current versus base level of the front gate pulse in SIMOX diodes with different lengths (after Ouisse et al).

In the inset of Figure 6.16(a), the CP current, I_{CP}, was measured by applying constant pulses on the front gate and varying V_{G2}. A sharp peak is observed in the region where the back interface is *depleted* [47]. This peak indicates that pulsing the front gate results in a scanning of the back surface potential which enables the pumping of some of the back interface traps.

Another interface coupling effect is that the values of the threshold voltage, flat band voltage and swept potential range at the front surface depend on the status (accumulation, depletion or inversion) of the back interface. This is the reason for the small difference existing between the CP currents measured for inversion and accumulation at the back interface. It might be envisaged to extract the concentration of back interface traps D_{it2} from the peak of I_{CP}, but this would require solving numerically the time-dependent Poisson equation across the film [48].

Figure 6.16(a) shows the modification of the "rectangle–like" CP curve as the back interface goes from inversion to depletion and accumulation. The shifts of the left and right edges are used to monitor the variations of V_{T_1} and V_{FB_1} versus V_{G_2}. The possibility to extract V_{T_1} easily, even for inversion at the opposite interface, is extremely useful because the standard extrapolation technique from $I_D(V_{G_1})$ MOSFET curves fails when both channels are activated. It is worth noting that, for the front interface traps to be accurately characterized, the back gate has to be maintained in strong inversion or accumulation.

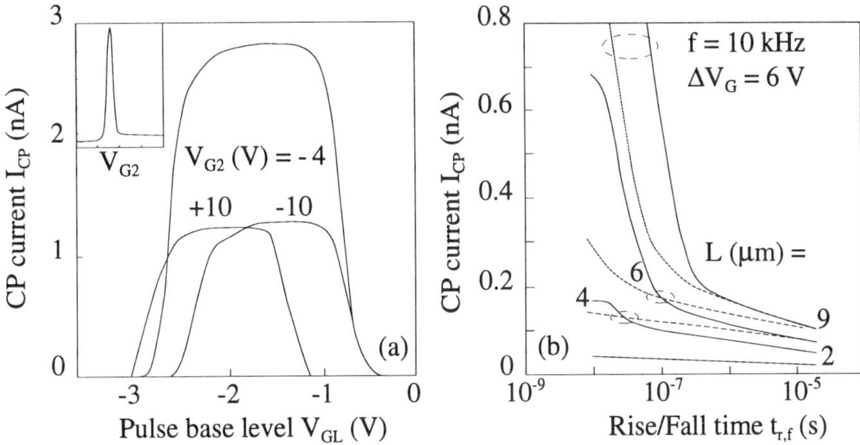

Figure 6.16: Front gate CP current versus (a) pulse base level for various back gate voltages, and (b) versus rise and fall times ($T_R = T_F$) for various diode lengths, with the back interface in inversion (——) or in accumulation (- - -), (after Ouisse et al).

Charge pumping depends on several *physical* time constants (carrier transit times for the formation of the accumulation and inversion channels, carrier lifetime, trapping and emission times) as well as on *experimental* time constants (period, rise and fall times). The experimental parameters have to be longer than the former ones, in order to prevent any source of parasitic recombination that would affect the evaluation of the recombination of trapped minority carriers with majority carriers. For instance, if all mobile minority carriers are not removed rapidly enough (during the fall of the pulse), before the arrival of majority carriers, an undesirable recombination occurs.

This parasitic component of I_{CP}, referred to as a "dimensional effect" [47], is responsible for an unacceptable overestimation of D_{it}, It obviously happens in long devices (the time for channel formation being proportional to L^2) or if the rise/fall times are too short. Since shorter the rise/fall times, wider the energy range of extracted interface trap profiles $D_{it}(E)$, it follows that the extension of energy profiles is limited by the occurrence of dimensional effects.

SOI structures are more subjected to dimensional effects because electron or hole reservoirs may exist at the back interface and interact with front channel carriers. A detailed analysis [47] demonstrates that parasitic recombination takes place essentially during the removal of either majority carriers (if V_{G2} is in inversion) or minority carriers (if V_{G2} is in accumulation). For instance, for $V_{G2} < 0$, the vertical field in the film assists the holes in leaving the surface and reaching the back P$^+$ reservoir or the P$^+$ terminal, whereas it is unfavorable for electron evacuation.

The range where the extraction of CP parameters is reliable is obtained by drawing $I_{CP}(\ln T_{R,F})$ curves which, according to the theory, must be linear. The onset of a non-linearity is evocative for the onset of dimensional effects (Figure 6.16(b)). The most critical case, $V_{G2} > 0$, corresponds to a difficulty in evacuating the holes which are less mobile. Dimensional effects are also revealed by the deformation of the CP peak (inset of Figure 6.16(a)) which would become highly asymmetric. However, Figure 6.16(b) shows that the rise/fall times may actually be small enough: $1\,\mu s$ for $L = 10\,\mu m$, $100\,ns$ for $L = 4\,\mu m$, $10\,ns$ for $L \leq 2\,\mu m$.

6.9   Low Frequency Noise

The analysis of the electrical noise is relevant in two respects: (i) it provides a practical limit for the operation of integrated circuits, and (ii) it represents a powerful tool for the identification and monitoring of oxide, semiconductor and interface defects. The noise is defined as the power spectral density $S_x(f)$ of a time-dependent, but stationary, random function $X(t)$, and it is given by the Fourier transform of the auto-correlation function of $X(t)$ [49]. In a semi-conductor device, the noise reflects the current fluctuations brought about by random variations in the carrier mobility and/or concentration. Each physical mechanism involved in current fluctuations is expressed by a specific random distribution (Gaussian, Poisson, ...) and has, in general, a distinct noise spectrum.

There is strong experimental evidence that the $1/f$ noise of MOS transistors originates from fluctuations in the carrier number. The basic mechanism is the carrier tunneling between the inversion channel and slow oxide traps N_t located at 1–2 nm from the Si–SiO$_2$ interface. The equivalent drain current noise is expressed as [50, 51]

$$\frac{S_{I_D}}{I_D^2} = \frac{q^4 \lambda_{ox}}{\alpha^2 kTLW} \times \frac{N_t}{(C_{ox} + C_d + qD_{it} + C_{inv})^2} \times \frac{1}{f} \qquad (6.42)$$

where λ_{ox} is an average tunneling length. The concentrations of slow traps N_t and interface traps D_{it} may be correlated by a first order approximation $D_{it} \simeq \lambda_{ox} N_t$. The inversion layer capacitance is $C_{inv} = (q/\alpha kT)Q_{inv}$, with $\alpha = 1$ in weak inversion and $\alpha = 2$ in strong inversion. In most experiments, a $1/f^\gamma$ dependence, with $0.8 \leq \gamma \leq 1.2$, is observed instead of a pure $1/f$ noise.

The advantage of Equation 6.42, also called Reimbold's equation, is to illustrate in detail the influence of the device parameters, from weak inversion to strong inversion, and from ohmic region to saturation region. In *weak inversion*, the variations of C_{inv} and C_d are negligible, so that the normalized noise factor S_{I_D}/I_D^2 is a constant, totally independent of drain or gate voltages. In *strong inversion*, C_{inv} becomes rapidly prevailing in Equation 6.42. A slight increase of the noise factor with V_D is observed due to the decrease in Q_{inv} along the channel (Figure 6.17(b)).

The typical noise curve, S_{I_D}/I_D^2 *vs.* V_G, shown in Figure 6.17(a), is composed of a "plateau" in weak inversion, followed by a sharp decrease, proportional to $(V_G - V_T)^{-2}$ in strong inversion. This ideal curve may be altered by inhomogeneous distributions of traps along the channel [52], in the energy gap [51], or within the oxide [53].

The *Random Telegraph Signal (RTS)* is attributed to individual carrier trapping at the Si–SiO$_2$ interface. In very small area devices, the trapping events are seldom and the trapping of one carrier can easily be detected in the time domain as a small pulse superimposed on the "normal" current value (also affected by other sources of noise). The duration of the pulse is nothing but the time constant of the trap. These pulses have rather sharp transitions and constant amplitude (Figure 6.18(a)). They correspond to the local modification of the flat-band voltage by one trapped carrier. It is shown that, for small enough fluctuations, the drain current pulse ΔI_D is independent of the trap area [54]

$$\frac{\Delta I_D}{I_D} = \frac{q}{LWC_{ox}}\frac{g_m}{I_D}\left(1 - \frac{d}{t_{ox}}\right) \tag{6.43}$$

where d is the distance of the trap from the interface. Not only, in very small devices, the trapping events become distinct but also does the amplitude of the fluctuation increase. In this respect, the back channel of SOI MOSFET's is an ideal tool for studying RTS properties because C_{ox_2} is very small and allows larger pulses to be obtained.

The power spectral density of RTS signals, induced by a single type of traps, is similar to that of G–R noise: plateau at low frequency followed by an $1/f^2$ decrease. When many traps with different time constants coexist, the superposition of their RTS contributions results in the conventional $1/f$ noise.

In early SIMOX material, annealed at low temperatures ($< 1250°C$), the dominant noises were $1/f$ and Lorentzian (with $1/f^2$ tail) [55]. Figure 6.17 shows the typical variations of the normalized $1/f$ noise factor S_I/I_D^2 with the gate and drain voltages. The transition from the weak inversion "plateau" to the strong inversion slope (Figure 6.17(a)) coincides with the threshold voltage. The density of slow traps is calculated from the plateau level using Equation 6.42[3]. It is found that the buried oxide contains, in general, one order of magnitude more slow traps than the front thermal oxide where $N_{t1} \leq 10^{18}$ cm^{-3}eV^{-1}.

---

[3] The capacitance term in the denominator of Equation 6.42 is easily calculated being the same as in the subthreshold swing (see Equation 6.25).

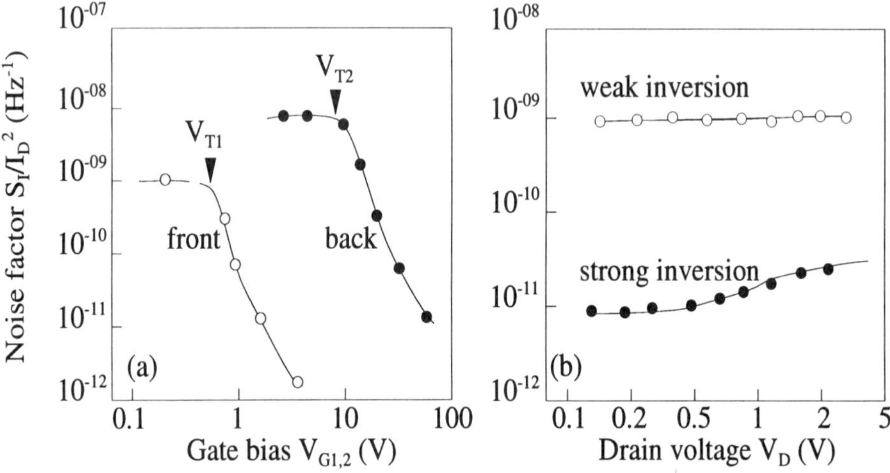

Figure 6.17: Normalized noise factor versus gate voltage (Fig.(a): oo front channel, •• back channel) and drain voltage (Fig.(b): oo weak inversion, •• strong inversion) in a SIMOX n–MOSFET.

In SIMOX structures annealed at high temperatures ($\geq 1300\,^{\circ}\mathrm{C}$), the $1/f$ noise is reduced and can be masked by a Lorentzian noise. When probing the back channel, an additional source of G–R noise was sometimes observed for special values of the back gate voltage. It was attributed to the activation of traps localized close to the buried oxide. Their energy level was deduced by computing the position of the Fermi level.

Typical RTS signals have been detected in small-area SIMOX transistors (Figure 6.18(a)). The pure RTS noise had a Lorentzian spectrum. Once the RTS signal was subtracted from the total current fluctuations, the residual noise had a white or $1/f$ spectrum [54].

In *depletion-mode* SOI MOSFET's, the two interface noise sources are completed by the noise generated in the neutral region of the film [56]. The drain voltage fluctuations, $S^{\star}_{V_D}$, measured in a partially depleted SIMOX N^+NN^+ transistor are shown in Figure 6.18(b). A very meaningful curve is obtained by varying the back gate bias while keeping the front interface in strong inversion. As long as the back interface is inverted too ($V_{G_2} < -20\,\mathrm{V}$), the noise is constant and reflects the unique contribution of the neutral film whose thickness is minimum (*i.e* maximum depletion regions). When V_{G_2} scans the weak inversion and depletion regions, the noise increases sharply due to the interactions between free carriers and buried oxide/interface traps. The noise peaks at the flat-band voltage. As V_{G_2} goes into accumulation, the interface traps are saturated and the measured noise decreases. A symmetrical behavior is obtained by varying V_{G_1} and maintaining the back interface strongly inverted.

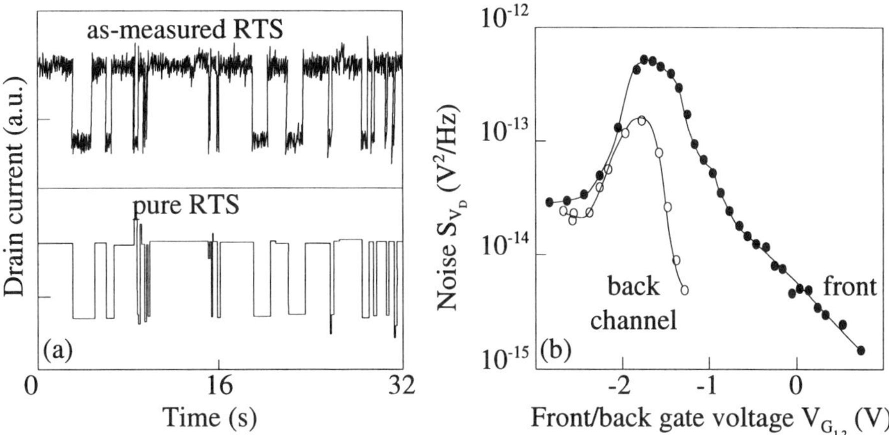

Figure 6.18: (a) As-measured and pure RTS signals in the time domain and (b) drain voltage noise (f=10 Hz, strong inversion) versus front and back gate voltages ($V_{G_2} = 10V_{G_1}$).

The interface coupling impacts on noise measurements in fully depleted SOI transistors in two main respects: *direct* noise generation at the opposite interface and *indirect* modification of the surface potential by the opposite gate bias. The front gate related noise now strongly depends on the back gate voltage. The shift of the noise peak position perfectly coincides with that of the extrapolated transistor threshold, here $V_{FB_1}(V_{G_2})$. As V_{G_2} becomes more negative, the back interface is gradually driven from accumulation to weak inversion and generates more noise.

In summary, the noise spectroscopy is an elegant method for discriminating and characterizing the defects located in different regions of the transistor. This only requires to "decouple" a given region from its surroundings, by appropriately biasing the front and back gates, and measure the noise generated solely by that source.

6.10 Drain Current Transient Technique

The carrier generation-recombination properties govern the performance of MOS and bipolar devices (leakage current of p-n junctions, refresh time of dynamic RAMs, floating body, bipolar gain, etc). Moreover, the carrier lifetime depends on residual crystalline defects and stands as a figure of merit for the quality of SOI films.

The well known Zerbst method [57] consists in applying a pulse on the gate of a MOS capacitor, to form a deep-depletion region. Equilibrium is reached through bulk and interface carrier generation processes. The monitoring of the capacitance transient provides the *generation lifetime* τ (which can substantially exceed the recombination lifetime) as well as the surface velocity S. Since capacitance measurements are not straightforward in thin SOI films, several modifications are needed: (i) MOS transistors will serve as test devices, (ii) the drain current will be monitored, and (iii) the dual gate configuration will be used to create deep depletion regions and initiate carrier generation. Discussed below are the variants of the technique for partially depleted accumulation-mode and inversion-mode SOI MOSFETs.

6.10.1 Partially Depleted Accumulation-Mode MOSFETs

We consider the N^+NN^+ transistor of Figure 6.19(a) biased in the ohmic region. Applying, at $t = 0$, a large negative voltage on the gate results in a very fast build-up of positive charge (deep depletion region) to balance the gate charge. The inversion charge starts forming, but it is a much slower process since no reservoir of holes is available. The deep depletion region controls the thickness of the drain current path. The holes supplied by thermal generation allow the deep depletion region to shrink and the current $I_D(t)$ to increase towards an equilibrium value I_∞ (Figure 6.19(a)) [58]. The transient variations of the inversion charge Q_{inv}, depletion charge Q_d, interface trap charge Q_{it}, and surface potential ψ_s are governed by the charge balance equation

$$Q_{inv} + Q_{it} = C_{ox}(V_G - V_{FB} - \psi_s) - Q_d \tag{6.44}$$

Expressing Q_d, ψ_s, and I_D in terms of depletion depth $w_d(t)$ gives

$$Q_d = qw_dN_D \qquad\qquad \psi_s = \frac{qw_d^2N_D}{2\epsilon_{si}} \tag{6.45}$$

$$I_D = \frac{q\mu N_D W V_D}{L}(t_{si} - w_d) = \frac{1}{K}(t_{si} - w_d) \tag{6.46}$$

The combination of Equations (6.44) and (6.45), yields the rate of change of the inversion and interface trap charges [59]

$$G = \frac{1}{q}\frac{d(Q_{inv} + Q_{it})}{dt} = -\frac{N_D C_{ox}}{2\epsilon_{si}}\left(1 + \frac{w_d}{\epsilon_{si}}\right)\frac{dw_d}{dt} \tag{6.47}$$

which is perfectly balanced by the total carrier generation rate

$$G = \frac{n_i(w_d(t) - w_\infty)}{\tau} + n_iS + n_iS_e\frac{(w_d(t) - w_\infty)(2L)}{WL} \tag{6.48}$$

The first term of Equation 6.48 stands for volume generation which is usually assumed to occur in the excess depletion region $(w_d - w_\infty)$, where w_∞ is the depletion depth at equilibrium [60]. The second contribution comes from the Si-SiO$_2$ interface and the last term involves the surface generation velocity S_e

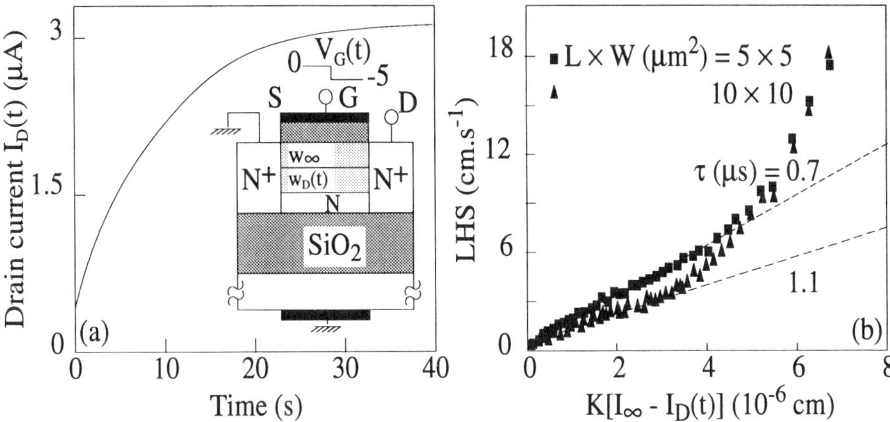

Figure 6.19: Accumulation-mode N^+NN^+ transistors pulsed in deep depletion ($V_{G_1} : 0 \rightarrow -5\,\text{V}$). (a) Drain current transient and (b) Zerbst-type analysis for transistors with different sizes (after Elewa et al).

on the edges of the silicon island. The active edge area, $2L(w_d - w_\infty)$, has been normalized with respect to the gate area. It is convenient to define an *effective* generation lifetime τ_{eff} by

$$\frac{1}{\tau_{eff}} = \frac{1}{\tau} + \frac{2S_e}{W} \tag{6.49}$$

Equating (6.47) and (6.48) results in a generic Zerbst-type relationship

$$LHS(t) = \frac{1}{\tau_{eff}} \times RHS(t) + S \tag{6.50}$$

The left-hand and right-hand sides are readily expressed, using Equation 6.46, in terms of drain current

$$LHS(t) = -\frac{N_D C_{ox}}{2n_i \epsilon_s} \frac{d}{dt}\left[t_{si} + \frac{\epsilon_{si}}{C_{ox}} - KI_D\right]^2$$
$$RHS(t) = K(I_\infty - I_D) \tag{6.51}$$

Instead of computing the current transient $I_D(t)$, Equation 6.50 is an elegant way of presenting the experimental data, LHS(t) *vs.* RHS(t), on a straight line (Figure 6.19(b)). The slope and intercept yield the values of τ_{eff} and S, respectively.

6.10.2 Partially Depleted Inversion–Mode MOSFETs

By contrast to accumulation-mode transistors, in N^+PN^+ structures, the drain and source reservoirs do supply minority carriers to form quasi-instantaneously

194

the inversion channel. Since there is no source of majority carriers (no P$^+$ contact), except the generation process, the experiment is now designed to induce a deficit of holes budget. The front gate is biased in strong inversion, whereas the back gate is pulsed from depletion (or accumulation) to strong accumulation.

The holes requested in the accumulation layer are released from the neutral region of the film. A deep depletion region forms under the gate and the film potential drops. To maintain front gate charge conservation (Equation 6.44), the excess depletion charge is compensated by a temporary drop in the inversion charge and current. The current relaxes back to equilibrium through carrier generation.

The front surface potential being almost constant in strong inversion ($\psi_{s_1} \simeq 2\phi_F$), it is the film potential $V_B(t)$ that governs the densities of inversion, depletion, and accumulation charges

$$
\begin{aligned}
Q_{inv} + Q_{it} &= -C_{ox_1}(V_{G_1} - V_{FB_1} - \psi_{s_1}) - Q_d \\
Q_d(t) &= -qw_dN_A = -\sqrt{2q\epsilon_{si}N_A(\psi_{s_1} - V_B(t))} \\
Q_{acc}(t) &= -C_{ox_2}(V_{G_2} - V_{FB_2} - V_B(t))
\end{aligned}
\tag{6.52}
$$

The charge balance of majority carriers is fulfilled by equating the rate of change of depletion and accumulation charges with Equation 6.48. Eliminating $V_B(t)$ again yields the Zerbst equation (6.50), where the right-hand side RHS(t) is the same as in Equation 6.51 and the left-hand side becomes

$$
LHS(t) = -\frac{N_A C_{ox_2}}{2n_i\epsilon_{si}}\frac{d}{dt}\left[K(I_\infty - I_D) + \frac{\epsilon_{si}}{C_{ox_2}} + \sqrt{\frac{4\epsilon_{si}\phi_F}{qN_A}}\right]^2
\tag{6.53}
$$

In some particular situations [61], the variation of the accumulation charge may be safely neglected, and the current has a simple exponential transient [62]

$$
I_\infty - I_D = I_\infty e^{-t/\tau_R}
\tag{6.54}
$$

where the recovery time is $\tau_R = \tau N_A/n_i$. This explains the difference by many orders of magnitude between the time constant of the generation process and the experimental transient of the drain current.

For large pulses, the deep depletion charge may be enough to temporarily suppress the drain current (for instance, when full depletion is achieved in accumulation-mode transistors). During a time t_{off}, freshly generated holes contribute to the relaxation of the front surface potential, which is swept from depletion and weak inversion to strong inversion. The off-state period can be calculated using the above equations [61]. A simpler method is to equate the positive charge generated during t_{off} with the gate charge in excess over that needed to cancel the current [63]. Of course, the standard Zerbst analysis still holds for $t > t_{off}$.

In films that are fully depleted at equilibrium, the concept of deep depletion and the expression $w_d(t)$ are meaningless. However, the dual gate pulsing technique can still be successful. Numerical computations of the generation rate show that the *"effective"* generation width corresponds to the film region where the electron and hole concentrations are smaller than the intrinsic carrier concentration [64]. Remark that as long as the two interfaces are depleted, they play equivalent roles in the generation process. The charge balance equation becomes

$$\frac{dQ_a}{dt} = - \left(1 + \frac{C_{ox_2}}{C_{ox_1}}\right) \frac{dQ_{inv}}{dt} = qn_i \left(\frac{w_{d,eff}}{\tau_{eff}} + S_{eff}\right) \qquad (6.55)$$

and can be rewritten in terms of $I_D(t)$.

The experiment is performed in obscurity, with low voltage applied on the drain (20–100 mV). Long channels are suitable because the extension of source and drain depletion regions is less significant. Junction leakage or BOX leakage currents may jeopardize the accuracy of drain current transients. These parasitic effects may be considered as negligible if the same set of generation parameters (τ, S) is obtained for several channel lengths.

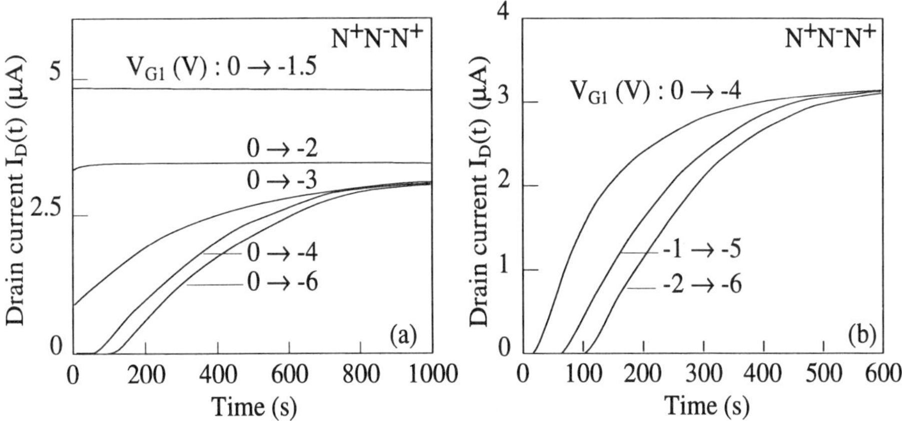

Figure 6.20: Typical transients in accumulation-mode N⁺NN⁺ transistors for different (a) bottom levels and (b) top levels of the pulse.

The measurements are performed with a HP–4145 system and a typical transient is shown in Figure 6.19 for accumulation-mode SOI MOSFETs. The data has numerically been differentiated according to Equation 6.50. As is usual for Zerbst plots, straight lines are obtained except at the beginning of the transient (right side of the Zerbst curves of Figure 6.19(b)). This is due to the more intense surface generation rate when the interface is depleted.

In Figure 6.20(a), the front gate of N⁺NN⁺ transistors was pulsed from zero towards increasing negative values. No transient is observed for small steps in gate bias (-2 V), simply because the depletion depth is narrower than w_∞. The electron population is promptly adjusted by the N^+ contacts and there is no need for hole supply so far. For -3 V step, the deep depletion depth lies between w_∞ and t_{si} and the initial current is non-zero. For yet larger steps, the film becomes totally depleted, the current drops to zero, and the off-state region t_{off} increases. By changing the initial gate bias, while maintaining the same step amplitude, the curves $I_D(t)$ are shifted and t_{off} varies (Figure 6.20(b)). However, the transients still converge to the same saturation current. The dominant parameter is the bottom pulse level: if it is kept constant and only the top level is modified, the transients remain more or less the same.

The roles of front and back gate biases in partially depleted enhancement-mode transistors are illustrated in Figures 6.21. When the back gate voltage step is increased, the current variation range is more pronounced, the off-state period is longer, but the steady-state current does not change (Figure 6.21(a)). Only in fully depleted MOSFETs, may I_∞ be modified by the pulse amplitude via interface coupling (Figure 6.21(b)); the bottom of the back gate pulse acts on the threshold voltage of the front channel. But, once the bottom of the pulse has reached strong accumulation, a further increase of the pulse will not modify any more V_{T_1} and I_∞ (Figure 6.21(c)).

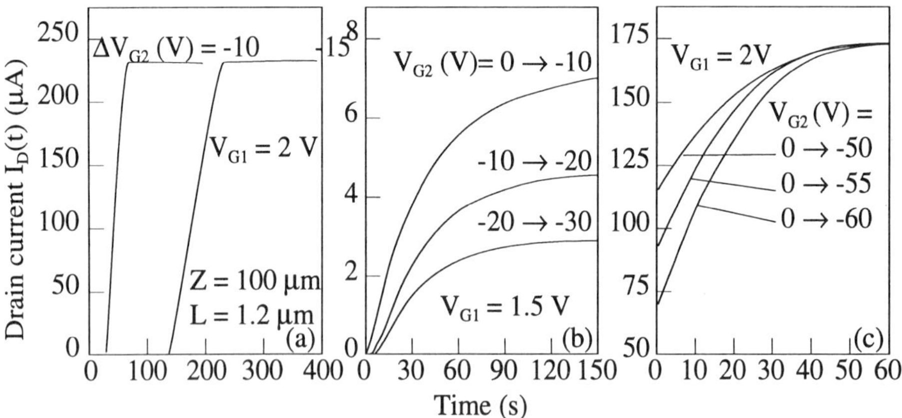

Figure 6.21: Current transients in enhancement-mode N^+PN^+ transistors pulsed with various back gate steps at fixed front gate bias: (a) partially depleted SIMOX, (b) fully depleted SIMOX pulsed in weak accumulation, (c) fully depleted wafer bonding MOSFETs pulsed in strong accumulation.

Increasing the front gate bias, the steady-state current becomes larger and the off-state region may be suppressed. This cancels or reduces the transition through the weak inversion regime and makes the strong inversion analysis with Equation 6.53 more accurate. For intentionally selected low gate biases, the weak inversion model may successfully be applied, leading to same lifetime values [65].

Measurements performed on various SOI materials reveal that the generation lifetime is very sensitive to the crystalline quality of the film. In early SIMOX films, annealed at low temperature, the carrier lifetime was very poor (10–100 ns). High temperature annealing above 1300 °C was remarkably efficient and improved the lifetime by three orders of magnitude (10–100 μs). So far, lifetime in excess of 100 μs has only been achieved in multiple-implanted SIMOX [61].

The surface generation velocity, currently deduced from Zerbst transients, is extremely small (0.01–0.1 cm/s) and does not reflect the actual generation rate at a *depleted* interface. It was found from weak inversion analysis that the surface generation velocity decreases by 2–3 orders of magnitude from weak to strong inversion [65]. The interface generation velocity on the silicon island sidewalls is much larger than at the front interface. It is obtained by comparing the effective lifetimes measured in normal and edgeless devices, or in transistors with various aspect ratios (see Equation 6.49).

The roles of the two gates can of course be interchanged for probing different regions of the film. Note that the transient duration can be much smaller, due to the difference in the capacitances of the gate oxide and buried oxide in Equations (6.51) and (6.53) without necessarily implying an inhomogeneity of the film quality.

The depth profiling of carrier lifetime is feasible by taking the derivative of the Zerbst curve with respect to w_d. The slopes of the tangents are simply the local values $\tau_{eff}^{-1}(w_d)$ at each point situated beyond w_∞.

In summary, the drain current transients are exclusive SOI techniques. They were imagined to exploit the special dual-gate configuration of SOI transistors, hence there is no chance for corresponding methods to exist in bulk Si.

6.11   Concluding Remarks

Several methods especially conceived for SOI or imported from bulk silicon have been revisited in the context of SOI materials and MOS transistors. Transport experiments in as-grown SOI wafers are rendered more difficult by the film thinness and full depletion condition. The Ψ–MOSFET appears to be a very unique and attractive technique for in-situ material evaluation.

As far as the device-based characterization methods are concerned, the adaptation of their experimental set-up and basic theory to thin SOI transistors is straightforward only if the back interface is maintained in accumulation. Otherwise, unusual and more complex situations are reached, as compared to the conventional curves obtained in bulk MOSFETs. Full depletion is the condition for interface coupling effects to occur, in other words the apparent carrier transport and generation properties at one interface can be controlled by the opposite gate.

None of the above characterization techniques can negate the use of the others. Each of them is focused on a particular feature and presents distinct merits and weaknesses. Together they can provide a comprehensive electrical image of SOI structures.

In this chapter, we have reviewed the most relevant characterization tools for SOI. Other techniques, as DLTS, PICTS, etc, may prove useful for particular evaluation purposes. Thanks to the development of appropriate characterization procedures, the SOI technology has reached a reasonable maturity. It is expected that SOI will soon compete with bulk silicon in the commercial arena of advanced intergrated circuits.

Bibliography

[1] S. Cristoloveanu and S.S. Li, *"Electrical Characterization of SOI Materials and Devices,"* Kluwer, Norwell, 1995.

[2] J. Margail, I. Stoemenos, C. Jaussaud, M. Dupuy, P. Martin, B. Blanchard, and M. Bruel, "Structural characterization of SIMOX structures," in *"Energy Beam Solid Interactions and Transient Thermal Processing,"* V.T. Nguyen and A.G. Cullis eds., Les Editions de Physique, Les Ulis, France, p. 519, 1986.

[3] S. Cristoloveanu, "Oxygen-related activity and other specific electrical properties of SIMOX," *Vacuum*, vol. 42, p. 371, 1991.

[4] J. Stoemenos, "Microstructure of SIMOX buried oxide, mechanisms of defect formation and related reliability issues," *Microelectronic Eng.*, vol. 22, p. 307, 1993.

[5] S. Cristoloveanu, "A review of the electrical properties of SIMOX substrates and their impact on device performance," *J. Electrochem. Soc.*, vol. 138, p. 3131, 1991.

[6] J.B. Lasky, "Wafer bonding for silicon-on-insulator technology," *Appl. Phys. Lett.*, vol. 48, p. 78, 1986.

[7] K. Mitani and U.M. Gösele, "Wafer bonding technology for silicon-on-insulator applications: a review," *J. Electronic Materials*, vol. 21, p. 669, 1992.

[8] P.B. Mumola, G.J. Gardopee, P.J. Clapis, C.B. Zarowin, L.D. Bollinger, and A.M. Ledger, "Plasma thinned SOI bonded wafers," *IEEE Int. SOI Conf. Proc.*, p. 152, 1992.

[9] W.P. Maszara, G. Goetz, A. Caviglia, and J. McKitterick, "Bonding of silicon wafers for silicon-on-insulator," *J. Appl. Phys.*, vol. 64, p. 4943, 1988.

[10] M. Haond, D.P. Vu, A.M. Aguirre, and S. Perret, "Electrical performances of devices made on SOI films obtained by lamp ZMR," *IEEE Circuits and Devices*, vol. 3, p. 27, 1987.

[11] B.Y. Tsaur, "Zone-melting-recrystal silicon-on-insulator technology," *IEEE Circuits and Devices*, vol. 3, p. 13, 1987.

[12] J.A. Knapp, "Silicon-on-insulator structures formed by aline-source electron beam: experiment and theory," *J. Appl. Phys.*, vol. 58, p. 2584, 1985.

[13] S.S. Tsao, "Porous silicon techniques for SOI structures," *IEEE Circuits and Devices*, p. 3, 1987.

[14] G. Bomchil, A. Halimaoui, and R. Herino, "Porous silicon: the material and its applications to SOI technologies," *Microelectronic Eng.*, vol. 8, p. 293, 1988.

[15] G.W. Cullen and C.C. Wang, *"Heteroepitaxial Semiconductors for Electronic Devices"*, Springer, Berlin, 1978.

[16] I. Golecki, "The current status of silicon-on-sapphire and other heteroepitaxial silicon-on-insulator technologies," in *"Comparison of Thin Film Transistor and SOI Technologies"*, H.W. Lam and H.J. Thompson eds., North-Holland, Amsterdam, p. 3, 1984.

[17] S. Cristoloveanu, "Silicon films on sapphire," *Rep. Prog. in Phys.*, vol. 50, p. 327, 1987.

[18] A.C. Ipri, "The properties of silicon-on-sapphire substrates, devices and integrated circuits," in *"Applied Solid State Science: Silicon Integrated Circuits"*, D. Kahng ed., Academic, New York, p. 253, 1981.

[19] S. Cristoloveanu and S. Williams, "Point-contact pseudo-MOSFET for in-situ characterization of as-grown silicon-on-insulator wafers," *IEEE Electron Device Lett.*, vol. 13, p. 102, 1992.

[20] S.T. Liu, P.S. Fechner, and R.L. Roisen, "Fast turn characterization of SIMOX wafers," in *IEEE SOS/SOI Technology Conf. Proc.*, p. 61, 1990.

[21] S. Williams, S. Cristoloveanu, and G. Campisi, "Point-contact pseudo-metal/oxide/semiconductor transistor in as-grown silicon on insulator wafers," *Materials Science and Eng.*, vol. B12, p. 191, 1992.

[22] H.S. Chen and S.S. Li, "A model for analyzing the interface properties of a semiconductor-insulator-semiconductor structure-I: Capacitance and conductance techniques," *IEEE Trans. Electron Dev.*, vol. 39, p. 1740, 1992.

[23] H.S. Chen and S.S. Li, "A model for analyzing the interface properties of a semiconductor-insulator-semiconductor structure-II: Transient capacitance technique," *IEEE Trans. Electron Dev.*, vol. 39, p. 1747, 1992.

[24] F.T. Brady, S.S. Li, D.E. Burk, and W.A. Krull, "Determination of the fixed oxide charge and interface trap densities for buried oxide layers formed by oxygen implantation," *Appl. Phys. Lett.*, vol. 14, p. 886, 1988.

[25] K. Nagai, T. Sekigawa, and Y. Hayashi, "Capacitance-voltage characteristics of semiconductor-insulator-semiconductor (SIS) structure," *Solid St. Electron.*, vol. 28, p. 789, 1985.

[26] E.H. Nicollian and A. Goetzberger, "Electrical properties as determined by the metal-insulator-silicon conductance technique," *Bell Syst. Technol. J.* vol. 46, p. 1056, 1967.

[27] S. Cristoloveanu and D. Ioannou, "Adjustable confinement of the electron gas in dual-gate silicon-on-insulator MOSFETs," *Superlattices and Microstructures*, vol. 8, p. 131, 1990.

[28] F. Balestra, S. Cristoloveanu, M. Benachir, J. Brini, and T. Elewa, "Double-gate silicon on insulator transistor with volume inversion: a new device with greatly enhanced performance," *IEEE Electron Device Lett.*, vol. 8, p. 410, 1987.

[29] D. Hisamoto, T. Kaga, and E. Takeda, "Impact of the vertical SOI 'DELTA' structure on planar device technology," *IEEE Trans. Electron Dev.*, vol. 38, p. 1419, 1991.

[30] J-P. Colinge, *Silicon-on-Insulator Technology: Materials to VLSI*, Kluwer, Boston, 1991.

[31] J.P. Colinge, X. Baie, and V. Bayot, "Evidence of two-dimensional carrier confinement in thin n-channel SOI gate-all-around (GAA) devices," *IEEE Electron Device Lett.*, vol. 15, p. 193, 1994.

[32] T. Ouisse, G. Ghibaudo, J. Brini, S. Cristoloveanu, and G. Borel, "Investigation of floating body effects in silicon-on-insulator metal-oxide-semiconductor field-effect transistors," *J. Appl. Phys.*, vol. 70, p. 3912, 1991.

[33] G.A. Armstrong, J.R. Davis, and A. Doyle, "Characterization of bipolar snapback and breakdown voltage in thin film SOI transistors by two-dimensional simulation," *IEEE Trans. Electron Devices*, vol. 38, p. 328, 1991.

[34] O. Faynot, A-J. Auberton-Hervé, and S. Cristoloveanu, "Investigation of the influence of the film thickness in accumulation-mode fully-depleted SIMOX MOSFETs," *ESSDERC'92 Conf. Proc.*, Elsevier, Amsterdam, p. 807, 1992.

[35] L.J. McDaid, S. Hall, P.H. Mellor, W. Eccleston, and J.C. Alderman, "Explanation of negative differential resistance in SOI MOSFETs," *ESSDERC'89 Proc. Conf.*, Springer, Berlin, p. 885, 1989.

[36] L.T. Su, K.E. Goodson, D.A. Antoniadis, M.I. Flik, and J.E. Chung, " Measurement and modeling of self-heating effects in SOI NMOSFETs," *IEDM'92 Conf. Proc.*, 1992.

[37] H-K. Lim and J.G. Fossum, "Threshold voltage of thin-film silicon-on-insulator (SOI) MOSFETs," *IEEE Trans. Electron Devices*, vol. ED–30, p. 1244, 1983.

[38] Y. Ohmura, S. Nakashima, K. Izumi, and T. Ishii, "0.1-μm-gate, ultrathin film CMOS devices using SIMOX substrate with 80-nm-thick buried oxide layer," *Proc. IEDM'91*, p. 675, 1991.

[39] B. Mazhari, S. Cristoloveanu, D.E. Ioannou, and A.L. Caviglia, "Properties of ultra-thin wafer-bonded silicon on insulator MOSFETs," *IEEE Trans. Electron Devices*, vol. ED–38, p. 1289, 1991.

[40] D.J. Wouters, J–P. Colinge, and H.E. Maes, "Subthreshold slope in thin-film SOI MOSFETs," *IEEE Trans. Electron Devices*, vol. ED–37, p. 2022, 1990.

[41] T. Ouisse, S. Cristoloveanu, and G. Borel, "Influence of series resistances and interface coupling on the transconductance of fully-depleted silicon-on-insulator MOSFETs," *Solid-State Electron.*, vol. 35, p. 141, 1992.

[42] S. Cristoloveanu, J–H. Lee, and A. Chovet, "Effets galvanomagnétiques dans les semiconducteurs anisotropes inhomogènes; application à la caractérisation des films de silicium sur saphir," *Rev. Phys. Appl.*, vol. 15, p. 725, 1980.

[43] A.C. Ipri, "Electrical properties of silicon films on sapphire using the MOS Hall technique," *J. Appl. Phys.*, vol. 43, p. 2770, 1972.

[44] G. Groeseneken, H.E. Maes, N. Beltran, and R.F. de Keersmaecker, "A reliable approach to charge pumping measurements in MOS transistors," *IEEE Trans. Electron Devices*, vol.ED–31, p. 42, 1984.

[45] T. Elewa, H. Haddara, S. Cristoloveanu, and M. Bruel, "Charge pumping in silicon on insulator structures using P–I–N diodes," in *Proc. 18th ESSDERC Conf.* (Montpellier, France), *J. Physique*, vol. 49, suppl. 9, p. C4-137, 1988.

[46] D.J. Wouters, M. Tack, G.V. Groeseneken, H. Maes, and C.L. Claeys, "Characterization of front and back Si–SiO$_2$ interfaces in thick and thin film silicon on insulator structures by the charge pumping technique," *IEEE Trans. Electron Devices*, vol.ED–36, p. 1746, 1989.

[47] T. Ouisse, S. Cristoloveanu, T. Elewa, H. Haddara, G. Borel, and D. Ioannou, "Adaptation of the charge pumping technique to gated p–i–n diodes fabricated on silicon on insulator," *IEEE Trans. Electron Devices*, vol. 38, p.1432, 1991.

[48] F. Hofmann and W. Hansch, "The charge pumping method: experiment and complete simulation," *J. Appl. Phys.*, vol. 66, p. 3092, 1989.

[49] A. Chovet and P. Viktorovitch, "Le bruit électrique," *L'Onde électrique*, vol. 57, p. 699 and 773, 1977, and vol. 58, p. 69, 1978.

[50] G. Ghibaudo, "A simple derivation of Reimbold's drain current spectrum formula for flicker noise in MOSFETs," *Solid–State Electron.*, vol. 30, p. 1037, 1987.

[51] G. Reimbold, "Modified $1/f$ trapping noise theory and experiments in MOS transistors biased from weak to strong inversion. Influence of interface states," *IEEE Trans. Electron Devices*, vol. ED–31, p. 1190, 1984.

[52] B. Boukriss, H. Haddara, S. Cristoloveanu, and A. Chovet, "Modeling of the $1/f$ noise overshot in short channel MOSFETs locally degraded by hot carrier injection," *IEEE Electron Device Letts.*, vol. 10, p. 433, 1989.

[53] Z.H. Fang, "Low-frequency pseudo generation-recombination noise of MOSFETs stressed by channel hot electrons in weak inversion," *IEEE Trans. Electron Devices*, vol. 33, p. 516, 1986.

[54] O. Roux dit Buisson, G. Ghibaudo, J. Brini, and T. Ouisse, "Measurements and analysis of random telegraph signals in small area SOI MOSFETs," in *Silicon–On–Insulator Technology and Devices* (W.E. Bailey ed.), The Electrochemical Soc., vol. 92–13, p. 191, 1992.

[55] A. Chovet, B. Boukriss, T. Elewa, and S. Cristoloveanu, "Low frequency noise of front channel and back channel MOSFETs fabricated on silicon on insulator SIMOX substrates," in *Noise in Physical Systems* (C.M. Van Vliet ed.), World Scientific, Teaneck, New Jersey, p. 457, 1987.

[56] T. Elewa, B. Boukriss, H.S. Haddara, A. Chovet, and S. Cristoloveanu, "Low-frequency noise in depletion-mode SIMOX MOS transistors," *IEEE Trans. Electron Devices*, vol. ED–38, p. 323, 1991.

[57] M. Zerbst, "Relaxation effects at semiconductor-insulator interfaces," *Z. Angew. Phys.*, vol. 22, p. 30, 1966.

[58] D.P. Vu and J.C. Pfister, "Determination of minority carrier generation lifetime in beam-recrystallized silicon-on-insulator by using a depletion-mode transistor," *Appl. Phys. Letts.*, vol. 47, p. 950, 1985.

[59] T. Elewa, H. Haddara, and S. Cristoloveanu, "Interface properties and recombination mechanisms in SIMOX structures," in *The Physics and Technology of Amorphous SiO_2*, R.A.B. Devine ed, Plenum, New York, p. 553, 1988.

[60] R.F. Pierret, "Rapid interpretation of the MOS–C C–t transient," *IEEE Trans. Electron Devices*, vol. 25, p. 1157, 1978.

[61] D.E. Ioannou, S. Cristoloveanu, M. Mukherjee, and B. Mazhari, "Characterization of carrier generation in enhancement-mode SOI MOSFETs," *IEEE Electron Device Letts.*, vol. 11, p. 409, 1990.

[62] P.W. Barth and J.B. Angell, "A dual-gate deep-depletion technique for generation lifetime measurements," *IEEE Trans. Electron Devices*, vol. 27, p. 2252, 1980.

[63] T.I. Kamins, "Minority-carrier lifetime in dielectrically isolated single-crystal silicon films defined by electrochemical etching," *Solid State Electon.*, vol. 17, p. 675, 1974.

[64] N. Yasuda, K. Taniguchi, C. Hamaguchi, Y. Yamaguchi, and T. Nishimura, "New carrier lifetime measurement method for fully depleted SOI MOSFETs," *IEEE Trans. Electron Devices*, vol. 39, p. 1197, 1992.

[65] A. Ionescu and S. Cristoloveanu, "Carrier generation in thin SIMOX films by deep-depletion pulsing of MOS transistors," *Nucl. Instr. and Methods in Phys. Res.*, vol. B84, p. 265, 1994.

7

Modern Analog IC Characterization Techniques

Hing-Yan To and Mohamed Ismail
Ohio State University
Ohio, USA

7.1 Introduction

The integrated circuit technology used in very large scale integrated (VLSI) circuit fabrication of digital circuits has been employed to realize purely analog circuits or mixed signal (digital and analog) circuits. In contrast to the digital counterpart, high precision analog circuit performance is strongly dependent on the accuracy of the element matching. Typical elements that are used in analog CMOS design are PMOS/NMOS transistors, MOS capacitors and resistors. Therefore, the characterization and modeling of these elements are of paritcular importance for CMOS analog circuit design. The basic characterization techniques are covered in other publications such as [1], [2] and [3]. This chapter discusses a number of characterization techniques which address the special concerns of analog circuit designers such as transistor matching, resistance matching and base spreading resistance extraction for BJT.

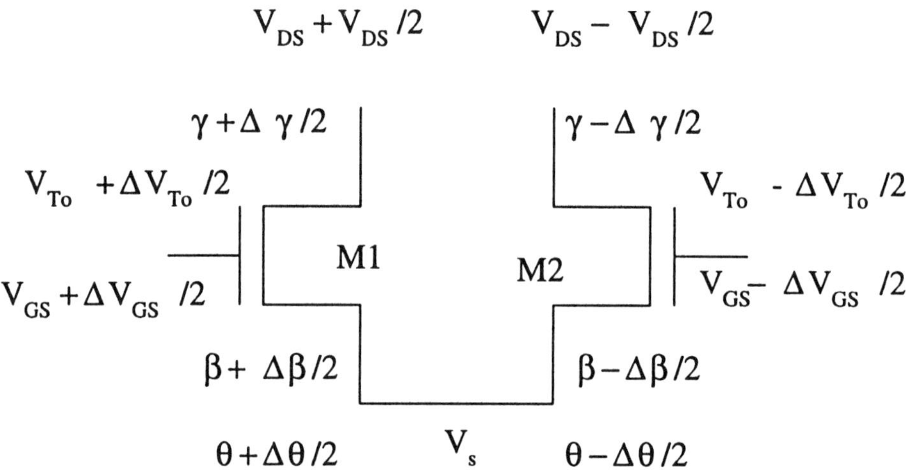

Figure 7.1: General Mismatched MOS Transistor Pair

7.2 Random Mismatch in MOS Transistors

The effect of transistor parameter variation on circuit performance is critical, especially for precision analog integrated circuit design. As the minimum feature size decreases, it is expected that the random process variations, which result in transistor mismatch, will become more significant. Transistor mismatch can be defined as the unequal drain currents of a pair of transistors which are identically designed, processed and under identical bias conditions. Mismatch [4] is a time-independent random process. Figure 7.1 shows a general pair of NMOS transistors. The parameter mismatch between the two transistors in the threshold voltage (V_T), the body-effect parameter (γ), the current factor (β), and the mobility modulation (θ) are represented by ΔV_T, $\Delta\gamma$, $\Delta\beta$, and $\Delta\theta$, respectively. The gate and drain bias mismatches are ΔV_{GS} and ΔV_{DS}, respectively. A general parameter (P) of M1 is $P_o + \Delta P/2$ and that of M2 is $P_o - \Delta P/2$ where P_o is the nominal parameter value and ΔP is the mismatch. This notation does not lose any generality, because the mismatch values are random and they may take either positive or negative sign.

The extraction of parameter mismatch can be based on a relatively simple model for the drain current of a MOS transistor. That is,

$$I_D = \frac{\beta}{1 + \theta(V_{GS} - V_T)}[(V_{GS} - V_T)V_{DS} - \frac{1}{2}V_{DS}^2] \tag{7.1}$$

when the MOS transistor is operating in the linear region. In the stauration region we have

$$I_D = \frac{1}{2} \times \frac{\beta}{1 + \theta(V_{GS} - V_T)}(V_{GS} - V_T)^2 \qquad (7.2)$$

with $V_T = V_{To} + \gamma(\sqrt{V_{SB} + 2\phi_F} - \sqrt{2\phi_F})$ and $2\phi_F$ is the surface inversion potential. Expanding Equations 7.1 and 7.2 in Taylor series around the nominal bias point and neglecting higher order terms, then the fractional current mismatch, $\Delta I_D/I_D$, in the linear region is given by

$$\frac{\Delta I_D}{I_D} = \frac{-\Delta V_T}{V_{GS} - V_T} + \frac{\Delta\beta}{\beta} + \frac{\Delta V_{DS}}{V_{DS}} - \frac{(V_{GS} - V_T)\Delta\theta}{1 + \theta(V_{GS} - V_T)} + \frac{\Delta V_{GS}}{(V_{GS} - V_T)} \qquad (7.3)$$

and in the saturation region

$$\frac{\Delta I_D}{I_D} = \frac{-2\Delta V_T}{V_{GS} - V_T} + \frac{2\Delta V_{GS}}{V_{GS} - V_T} + \frac{\Delta\beta}{\beta} - \frac{(V_{GS} - V_T)\Delta\theta}{1 + \theta(V_{GS} - V_T)} \qquad (7.4)$$

where the threshold mismtach, ΔV_T, is a function of both the V_{To} and γ mismatches.

If both transistors are biased equally, then Equations 7.3 and 7.4 model the drain current mismatch as a function of the four parameter mismatches ΔV_{To}, $\Delta\gamma$, $\Delta\beta$, and $\Delta\theta$. For a randomly selected transistor pair, these mismatches are random variables, therefore, the drain current mismatch is also a random variable. A population of nominally identical transistor pairs with statistically large parameter mismatch variations will have correspondingly large variations of drain current mismatch.

7.2.1 Measurement Methodologies

An examination of Equations 7.3 and 7.4 reveals two important points, both of which were utilized in the mismatch measurement methodologies discussed here. It is obvious from (7.3) and (7.4) that a mismatch in gate bias, which might occur if the transistors were measured separately, has the same effect on the measured current mismatch as a threshold voltage mismatch. Therefore, an estimate of threshold voltage mismatch obtianed from sequential, rather than simultaneous, measurement of V_{T1} and V_{T2} may have significant error. Thus, in order to measure the mismatch of a transistor pair accurately and repeatably, it is necessary to bias and measure both transistors simultaneously, as shown in Figure 7.2. A typical common centroid layout of a pair of transistors is shown in Figure 7.3.

From Equation 7.3, it is also clear that a mismatch in drain bias during measurement can mask the true value of $\Delta\beta$. The effect of ΔV_{DS} can be minimized by using a large nominal V_{DS}, and may be completely eliminated in the ideal case by making measurements in saturation.

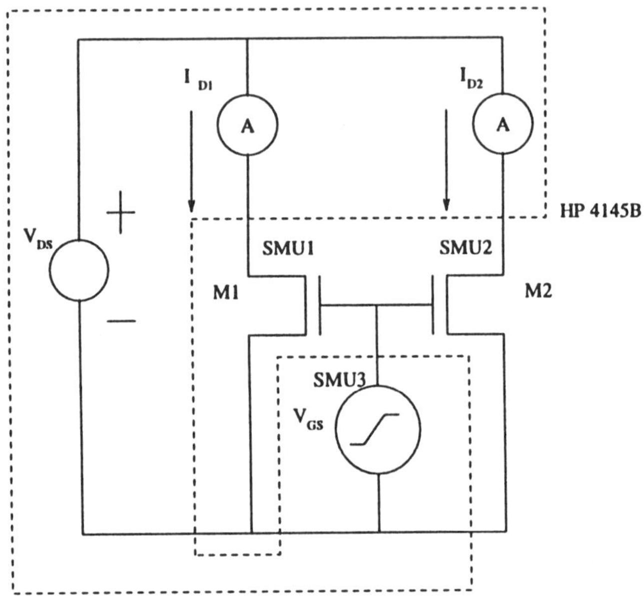

Figure 7.2: Simultaneous I_D Measurement of MOS transistor Pair (Schematic)

The parameter mismatches can be extracted using regression technique. Equation 7.4 is a linear function of the transistor parameter mismatches. That is,

$$z = C_1 X_1 + C_2 X_2 + C_3 \tag{7.5}$$

where z is $\Delta I_D / I_D$, X_1 is $-2/(V_{GS} - V_T)$, C_1 is ΔV_T, C_2 is $\Delta\theta$, X_2 is $-(V_{GS} - V_T)/(1 + \theta(V_{GS} - V_T))$ and C_3 is $\Delta\beta/\beta$. Based on the least square fit technique, the normal equations corresponding to Equation 7.5 are:

$$\begin{aligned}
\Sigma z &= C_1 \Sigma X_1 + C_2 \Sigma X_2 + n C_3 \\
\Sigma X_1 z &= C_1 \Sigma X_1^2 + C_2 \Sigma X_1 X_2 + C_3 \Sigma X_1 \\
\Sigma X_2 z &= C_1 \Sigma X_1 X_2 + C_2 \Sigma X_2^2 + C_3 \Sigma X_2
\end{aligned} \tag{7.6}$$

Different X_1 and X_2 are obtained at different biases for the transistor pair. By solving the normal equations, the transistor parameter mismatch for each pair of transistors can be determined.

Since mismatch is a random variable, the statisitcs of a parameter mismatch can be estimated by measuring a sample of transistor pairs from the same process. In addition, it is well known that the standard deviation of mismatch of a standard MOS parameter, $\sigma(\Delta P)$, in a large sample of matched transistors is inversely proportional to the square root of the active areas of the individual

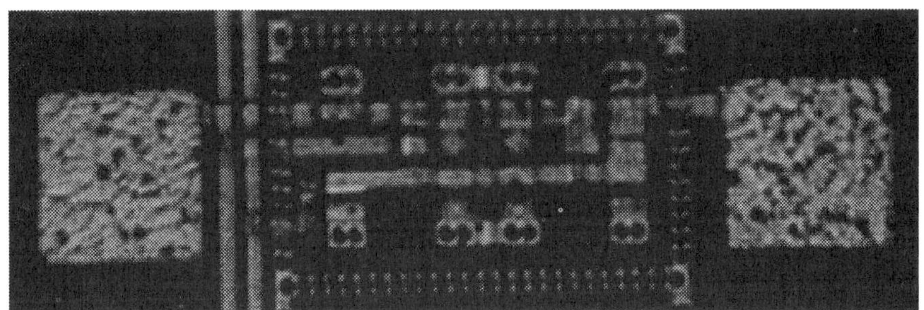

Figure 7.3: Actual Unit MOS Transistor Pair Test Structure

transistors. Mathematically speaking, this is in the notation of

$$\sigma(\Delta P) = \frac{\sqrt{a_P}}{\sqrt{WL}} \tag{7.7}$$

where a_P is a process-dependent fitting constant. The process fitting constants are estimated by measuring the standard deviations of parameter mismatch for different active area transistors. Typical values of the fitting constants and the correlation matrices for NMOS parameter mismatches and for PMOS parameter mismatches are shown in Tables 7.1, 7.2, and 7.3, respectively. The correlation bewteen the different mismatches is estimated by measuring the linear dependence on a scatter plot. It is important to preserve the correlation among the transistor parameters in statistical modeling. Since unrealistic combination of parameters can be created for simulation if correlation is not considered. Hence, the simulation is not accurate.

Table 7.1: The Area Fitting Constants

	NMOS ($\sqrt{a_P}$)	PMOS ($\sqrt{a_P}$)	Units
ΔV_T	31.0	33.2	$mV * \mu m$
$\Delta\beta/\beta$	10.0	4.89	$\% * \mu m$
$\Delta\theta/\theta$	10.4	6.19	$\% * \mu m$
$\Delta\gamma/\gamma$	6.10	2.86	$\% * \mu m$

Table 7.2: The Correlation Matrix for NMOS Parameter Mismatch

	ΔV_T	$\Delta\beta$	$\Delta\theta$	$\Delta\gamma$
ΔV_T	1.0	-0.345	-0.328	0.539
$\Delta\beta$	-0.345	1.0	0.971	-0.412
$\Delta\theta$	-0.328	0.971	1.0	-0.376
$\Delta\gamma$	0.539	-0.412	-0.376	1.0

Table 7.3: The Correlation Matrix for PMOS Parameter Mismatch

	ΔV_T	$\Delta\beta$	$\Delta\theta$	$\Delta\gamma$
ΔV_T	1.0	0.2646	0.275	0.183
$\Delta\beta$	0.246	1.0	0.978	-0.531
$\Delta\theta$	0.275	0.978	1.0	-0.601
$\Delta\gamma$	0.183	-0.531	-0.601	1.0

7.3 The Extraction of BJT Base Spreading Resistance

For high speed amplifiers, which utilize a bipolar common emitter input stage, most of the output noise can be attributed to the thermal noise generated in the base resistance of the input transistors. For collector currents larger than about 50μA, which would be typical for a high speed implementation, the noise generated in the base resistance of the input stage will be dominant.

This section will show how a direct measurement of the output noise spectrum can be used to determine the value of the base spreading resistance of the input devices [5].

7.3.1 Bipolar Noise and Performance Modeling

The bipolar noise properties are characterized by the shot noise, thermal noise, Flicker noise, and burst/popcorn noise.

The high frequency, white noise is generated by the shot noise and thermal noise. The shot noise results from the discrete quantized nature of the current flow across a semiconductor junction, while the thermal noise is generated by the random thermal motion of electrons in the lattice.

The low frequency noise is caused by the Flicker noise and burst noise. The Flicker noise is generated by fluctuations in current recombining at the base surface, while the burst noise originates from heavy metal contamination. For most bipolar devices, the low frequency noise sources are insignificant, and we are mainly concerned with the shot noise and thermal noise.

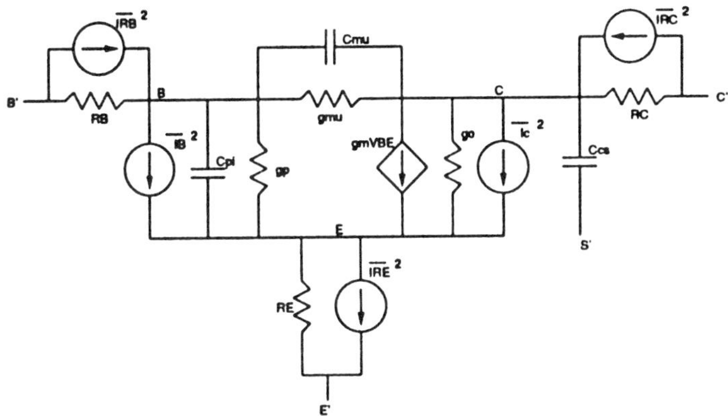

Figure 7.4: Bipolar small signal model with noise sources

Figure 7.4 shows the bipolar small signal model including noise current sources used by SPICE.

$$\overline{I_{RB}^2} = \frac{4KT}{R_B}\Delta f \qquad \overline{I_{RC}^2} = \frac{4KT}{R_C}\Delta f \qquad (7.8)$$

$$\overline{I_{RE}^2} = \frac{4KT}{2R_E}\Delta f + \frac{KF(I_B)^{AF}}{f}\Delta f$$

$$\overline{I_C^2} = 2qI_C\Delta f \qquad \overline{I_B^2} = 2qI_B\Delta f$$

The base resistance models the combined effect of the extrinsic base resistance seen between the intrinsic base and the base contact, and the large distributed resistance seen in the base layer. The extrinsic base resistance is modeled by a fixed separate entity, RBM, while the intrinsic resistance seen through the base region is modeled by RB, which is a function of the base current and the base resistance knee current (IRB) if specified in the SPICE model deck.

The typical procedure for extracting RB, is to first extract the extrinsic base resistance, RBM, and then extract the total base resistance for a given base current.

The extrinsic base resistance, RBM, is typically obtained by plotting the difference between the applied base emitter voltage and the measured open circuit collector-emitter voltage as a function of the inverse base current described by the following equation:

$$\frac{V_{BE} - V_{CE}}{I_B} = \frac{V_T}{I_B}\left[ln(\frac{\alpha_F I_B}{I_S}) - ln(\frac{1}{\alpha_R})\right] + R_B \qquad (7.9)$$

212

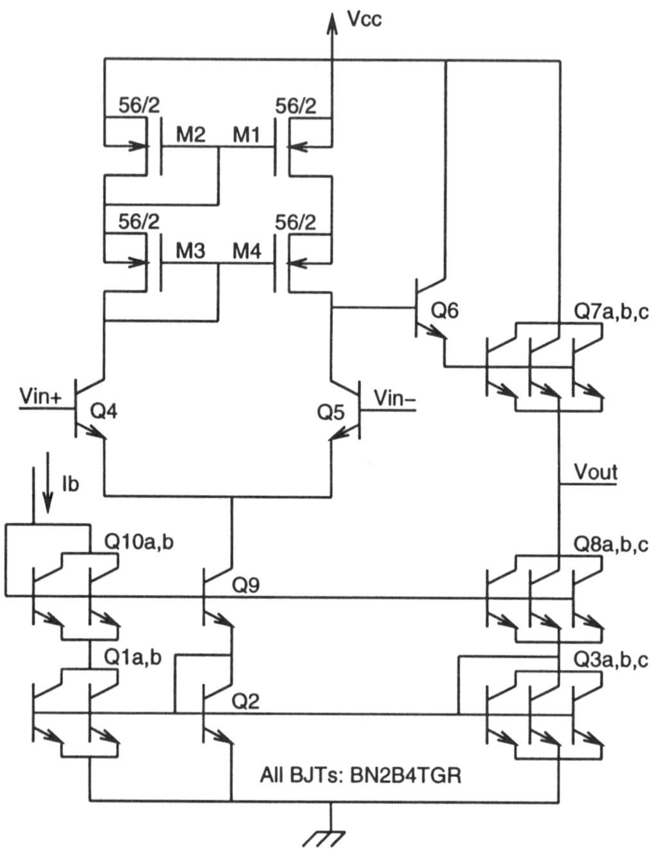

Figure 7.5: Schematic of a wide band amplifier used for the extraction of the base spreading resistance.

By extrapolating this curve, RBM can be obtained as the base resistance (R_B) for infinite base current.

A multitude of possible techniques exist for the extraction of the total base resistance, such as the common emitter impedance circle method, direct measurements on the Gummel plots, step response measurement techniques, phase cancellation techniques, time domain reflectory methods, noise figure degradation measurements, and finally direct measurement of the output noise spectrum of a common emitter amplifier [6]. In this section we will focus on the noise measurement technique.

Figure 7.5 shows the schematic of the wide band amplifier used for extraction of the base spreading resistance through noise measurement techniques. This amplifier was designed for maximum sensitivity to the noise introduced by the base resistance and for easy compensation of all other significant noise sources.

Bias current	100μA	500μA	1000μA
Noise Source	Noise Power Density, $[V^2/Hz]$		
Tot. PMOS load	1.67E-11	1.19E-12	4.94E-13
Tot. buffer	5.61E-13	2.77E-14	8.91E-15
Tot. biasing	7.59E-15	1.21E-15	2.71E-15
Tot. diff. pair	8.07E-11	2.85E-11	2.31E-11
Diff. breakdown			
RB (thrm)	2.14E-11	1.58E-11	1.52E-11
RE (thrm)	8.82E-13	8.74E-13	1.01E-12
RC (thrm)	1.00E-17	1.27E-17	1.30E-17
I_C (shot)	5.72E-11	1.17E-11	6.82E-12
I_B (shot)	2.10E-14	5.59E-14	9.03E-14
Tot. noise	9.80E-11	2.97E-11	2.36E-11
Uncomp. error	78.1%	46.8%	33.5%
Comp. error	17.6%	4.10%	2.14%

Table 7.4: Noise performance of BiCMOS amplifier

In order to generate a measurable quantity of white noise, we chose to use a cascode PMOS load for added gain. This implies that the effective gain is determined by the product of the bipolar output resistance and the bipolar transconductance assuming the conductance seen into the cascode PMOS load is negligible, and the gain should thus be independent of the bias current ($g_m r_o$ of a bipolar device is independent of collector current). Thus, by adjusting the biasing current, the resulting change in output noise will be caused only by the effect introduced by the different base current in the input stage, making extraction of IRB easy. By adjusting the bias current, the dominant pole location will move as an effect of the reduced output impedance, but as long as we measure the noise power spectrum below the open loop 3dB frequency, this will not complicate the measurement. Table 7.4 lists the most significant simulation results with respect to the noise performance of this amplifier when it was biased with 100μA, 500μA, and $1mA$.

In order to have the noise introduced by the base resistance of the input stage dominate, the collector current has to exceed 50μA, or similarly, the bias current has to exceed about 200μA.

Experimental measurements done with a bias current of 1mA yielded a low frequency gain of 58dB, and an open loop 3dB frequency of about 250kHz when loaded with 50pF. The Darlington output stage provided an output impedance of about 45Ω, for good matching with the measurement apparatus and easy drive of the high capacitance probe cables used ($\approx$50pF).

The biasing was provided by a modified Wilson current mirror for optimum matching between the collector currents in the different branches. Simulations

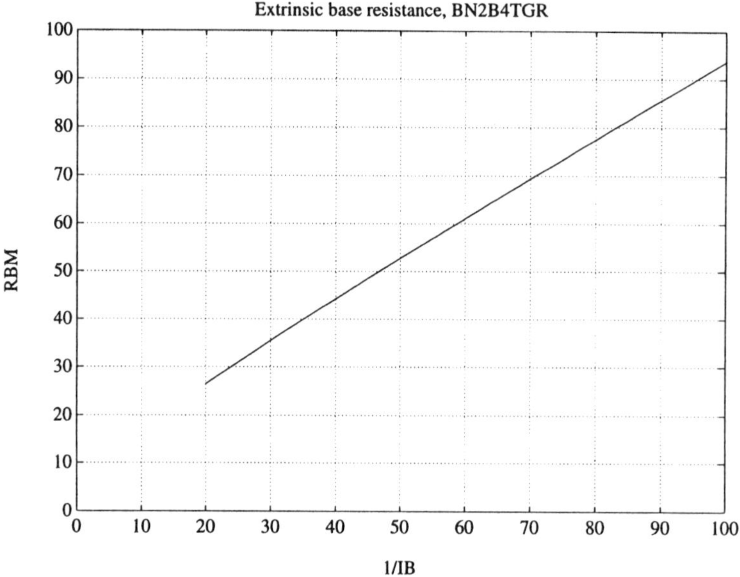

Figure 7.6: Extraction of the Extrinsic Base Resistance

yielded deviations less than 0.05% between the currents in the lower transistor pairs, and this system error will clearly be insignificant compared to random processing deviations.

Thus, from simple DC measurements on a single device for the current gain β and the measured external bias current, the internal collector current and base current can be determined. From Table 7.4 we find that the noise introduced by the combined base and emitter resistance of the input pair generates 66.5% of the total output noise measured on the amplifier, and if we compensate for the known collector and base shot noise, 97.86% of the total output noise can be accounted for. Since a common base and common emitter amplifier have the same voltage gain, it is impossible to separate the noise introduced by the emitter resistance from the noise introduced by the base resistance. The emitter resistance is however easily extracted from DC measurements of the open circuit collector-emitter voltage, and can thus be compensated for at a later point [6]. Since this process utilizes metal emitter contacts, and not poly emitters, the emitter resistance is also quite small.

7.3.2 Base Resistance Extraction

The extrinsic base resistance can be extracted using DC measurement techniques and an HP-4145 parameter analyzer. Figure 7.6 shows the measured base resistance as a function of the inverse base current.

Using Equation 7.9 the extrinsic, or minimum, base resistance value can be obtained by extrapolating this curve to $1/I_B$ equal to zero. Thus, for infinite

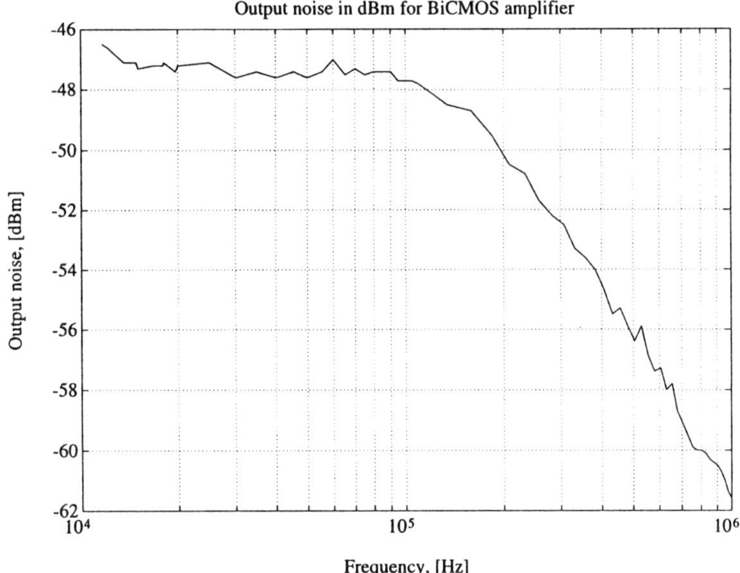

Figure 7.7: Output Noise Spectrum of Amplifier

base current, the base resistance is extracted to be 9.5Ω, which is RBM.

Figure 7.7 shows a measured output noise in dBm for the open circuited wide-band amplifier.

This plot was obtained by biasing the amplifier open circuited with the output quiescent voltage at mid rail. A single rail of 5V was used for these simulations and measurements. From the plot we find that the output noise starts to decrease at around 100kHz, which corresponds to the point where the open loop gain starts to decrease. At this point, noise introduced at the input of the amplifier will be amplified less, and the output noise spectrum starts to decrease.

The output noise spectrum generated by the bipolar input stage of the amplifier is given by the following equation

$$
\begin{aligned}
\frac{V_{out}^2}{\Delta f} &= 4kTRC + 2qI_C(RC + r_0)^2 \\
&+ A_0^2[8kT(RB + RE) + 4qI_B(RB + RE)^2]
\end{aligned}
\tag{7.10}
$$

where RC, RB, and RE are the respective ohmic stray resistances, r_0 is the small signal impedance looking into the collector of Q5, I_B and I_C are the quiescent base and collector currents, and A_0 is the gain of the amplifier. These terms should be recognized as the thermal noise introduced by the stray collector resistance of Q5, the shot noise introduced by the collector current in Q5, followed by the thermal noise generated by the combined base and emitter resistance of Q4 and Q5 and base shot noise generated by Q4 and Q5 both amplified by the open loop gain of the amplifier. It should be noted that only true ohmic resistances generate thermal noise, while shot noise is amplified by whatever small signal impedance it flows into. The input noise sources of both devices will generate noise at the output, while only the output noise sources of Q5 will affect the output noise.

We can solve Equation 7.10 for the sum of the base and emitter stray resistances once all other parameters are determined, and the output noise power spectral density measured. In the flat low frequency region, the gain of the amplifier was measured to be 58dB, or 795V/V. Using a spectrum analyzer, the noise power spectral density from 10kHz through 100kHz was determined to be $2.49E^{-6}V/\sqrt{Hz}$, or $6.24E^{-12}V^2/Hz$. In order to obtain the small signal impedance looking into the collector of Q5, we simply divide the open loop gain by the transconductance of Q5, which yields an output impedance of $84.343k\Omega$. The stray collector and emitter resistances were independently extracted from a single device to be 645Ω and 9Ω respectively. The current gain was also measured independently, and for 1mA of external bias current, the base and collector currents of the input transistors were determined to be $2.1\mu A$ and $241.64\mu A$. Solving (7.10), and referring all noise sources back to the input as resistors, we obtain the following resistance values from the measured data.

$$\frac{V_{out}^2/\Delta f}{4kT A_0^2} = \frac{6.24E^{-12}}{1.656E^{-20}795^2} = 596.20\Omega \tag{7.11}$$

$$RB + RE = 268.49\Omega$$

$$\frac{RC}{A_0^2} = \frac{641}{795^2} = 1.01m\Omega$$

$$\frac{2qI_C(r_o + RC)^2}{4kT A_0^2} = \frac{7.73E^{-23}(84343 + 641)^2}{1.656E^{-20} \times 795^2} = 53.36\Omega$$

$$\frac{4qI_B(RB + RE)^2}{4kT} = \frac{1.344E^{-24}(268.49)^2}{1.656E^{-20}} = 5.85\Omega$$

From these equations, we find that the total output noise can be represented by a 596.2Ω resistor sitting at the input terminal, and out of this a resistance of 536.98Ω is introduced by the combined base and emitter resistance of the input pair. Dividing by two, and subtracting the emitter resistance and the extrinsic base resistance, we find the internal base spreading resistance to be 249.99Ω.

This section discussed the technique to use a wide band BiCMOS amplifier to extract the base spreading resistance of bipolar devices. The extraction is done by measuring the output noise power spectral density.

7.4 Mismatch Characterization of BJT for Statistical CAD

With the increasing popularity of bipolar-complementary MOS (BiCMOS) technology, a statistical CAD compatible model for BJT mismatch, which can predict the mismatch across the entire bias range, becomes essential for statistical design and simulation [7]. Together with the existing statistical model for MOS transistors [8] [9], the statistical model for BJTs can be used to characterize BiCMOS designs.

It is determined [10] that the collector current mismatch of a BJT pair at any bias can be represented by two BJT parameter mismatches: I_s and λ. The mismatch in base current is neglected because it is usually much smaller than the collector current. The collector current I_C of a BJT in normal active region can be represented by the following equation [3] :

$$I_C = \frac{I_s}{1 + \lambda V_{BC}} e^{V_{BE}/V_T} \tag{7.12}$$

where $V_T = \frac{KT}{q}$, I_s is the saturation current of the transistor, and λ is the reciprocal of the Early voltage (V_A). The mismatches between two transistors in I_s, λ, V_{BC}, and V_{BE} are ΔI_s, $\Delta \lambda$, ΔV_{BC}, and ΔV_{BE}, respectively.

Let P be a general parameter, P_o be the nominal value of the parameter which is equal to $\frac{P_1 + P_2}{2}$ and ΔP be the parameter mismatch which is equal to $P_1 - P_2$. The collector currents, (I_{C1}) and (I_{C2}), in two presumably matched transistors can be expressed as $f_{I_C}(P_o + \frac{\Delta P}{2})$ and $f_{I_C}(P_o - \frac{\Delta P}{2})$, respectively. Since mismatch is a random variable, the assignment of sign is arbitrary. By Talyor expansion and neglecting the higher order terms, the fractional collector current mismatch is given by

$$\frac{\Delta I_C}{I_C} = \frac{\Delta I_s}{I_s} + \frac{\Delta V_{BE}}{V_T} - \Delta\lambda \frac{V_{BC}}{1 + \lambda V_{BC}} - \Delta V_{BC} \frac{\lambda}{1 + \lambda V_{BC}} \tag{7.13}$$

It is clear from Equation 7.13 that ΔV_{BE} can mask the effect of ΔI_s if measurements of different mismatches are done separately.

A cross-coupled BJT pair is used to extract the parameter mismatches. The configuration is shown in Figure 7.8. Transistors Q1 and Q2 (Q3 and Q4) are connected diagonally (Figure 7.8(b)) to form one composite transistor. The emitters of the two composite transistors are connected and so are their bases.

Also, their collectors are driven by the same V_{cc}. In doing so, ΔV_{BE} and ΔV_{BC} are zero. An HP 4145 parameter analyzer is used to measure the collector currents simultaneously. The mismatches $(\Delta I_s, \Delta \lambda)$ can be extracted by using the least square fit technique at different values of V_{BC}. A number of cross coupled BJTs for different emitter areas can be measured and the results are used to estimate the statistics of the mismatches. The typical plots of the standard deviation of I_s mismatch and λ mismatch are shown in Figure 7.9 and 7.10, respectively.

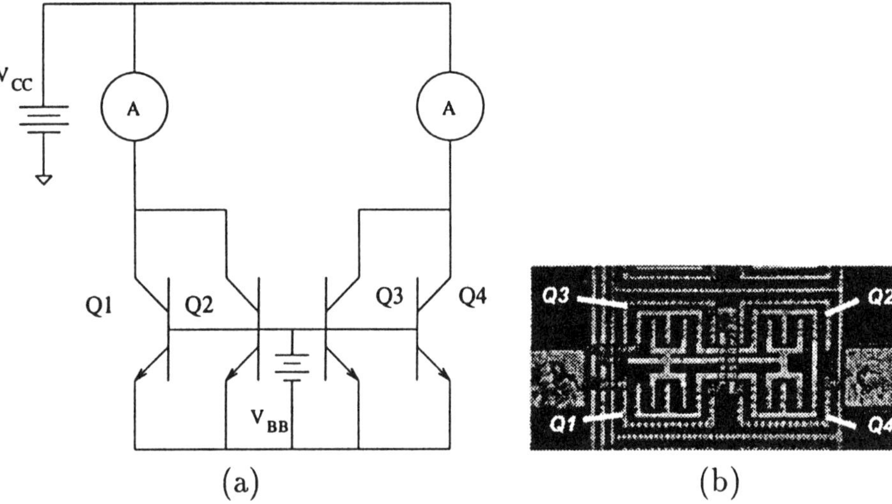

Figure 7.8: BJT Test Structure (a) Schematic (b) Actual Unit Test Structure

Table 7.5: Fitting Constants and Correlation Matrix

	$\sqrt{a_p}$	Units		ΔI_s	$\Delta\lambda$
$\Delta I_s/I_s$	1.10	$\%\mu m$	ΔI_s	1.0	0.55
$\Delta\lambda$	1.86×10^{-4}	$V^{-1}\mu m$	$\Delta\lambda$	0.55	1.0

The area fitting constants and correlation matrix obtained for a 2 μm n-well process are listed in Table 7.5.

7.5 Test Structure for Resistance Matching Properties

The measurement of accurate resistance values and the mismatch between resistors is critical in the design of analog integrated circuits [11]. For analog filter structures as well as digital/analog (D/A) and analog/digital (A/D) converters, the overall achievable accuracy is often limited by this factor [12].

Typically, the analog designer and the process engineer use different procedures and test structures to characterize resistive elements in a process. The process engineer would typically use Van der Pauw structures [13] to determine

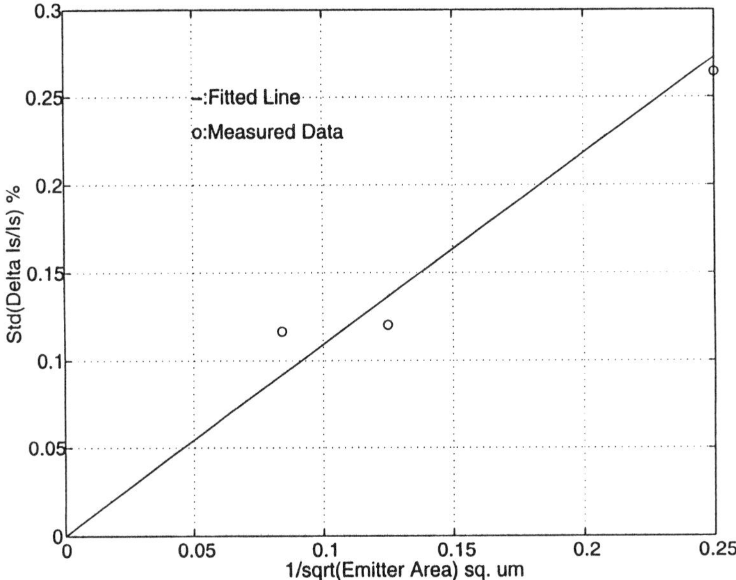

Figure 7.9: Standard Deviation of I_s Mismatch versus $(WL)^{-1/2}$

the sheet resistivity, a number of resistors with equal lengths to determine the lateral encroachment, ΔW, and a Kelvin contact resistance structure [14], or a simple contact chain in order to determine the combined contact and spreading resistance introduced by a given contact. On the other hand, the analog designer is more interested in how accurately a resistance ratio can be set; the actual absolute value of a resistor is usually of secondary importance. Thus the test structure used will typically contain resistors laid out symmetrically in a variety of large dimensions to determine the matching and scaling properties.

The test structure discussed here combines these two approaches. The analog designer can use this structure to statistically determine how closely two identical resistors can be matched, as well as how accurately two resistors can be matched to a given ratio. The process engineer can accurately determine the sheet resistivity, the encroachment, the end effects caused by current crowding and the ohmic contact resistance separately using the same structure. By performing a statistical analysis of all the extracted data, it can be determined where most of the mismatches and variations are introduced. The analog designer will also gain insight into how to design a given resistor ratio for optimum resistance matching by reducing the effect of the most unstable parameters.

Measuring all the parameters on the same structure makes it possible to detect local variations that are not totally random in any of the parameters and determine how variations of one parameter are linked with variations of another parameter.

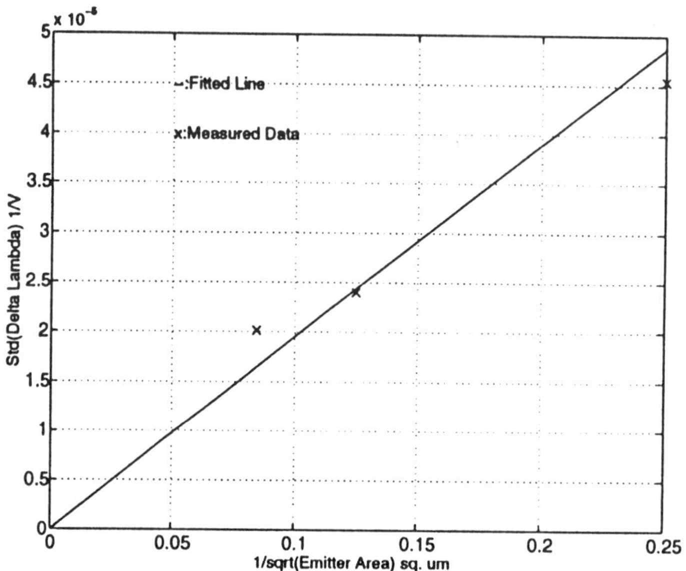

Figure 7.10: Standard Deviation of λ Mismatch versus $(WL)^{-1/2}$

7.5.1 Resistance Test Structure

For most processing lines and resistor structures, the resistance can be expressed by the following equation:

$$R = \rho \, \frac{L + \Delta L}{W + \Delta W} + 2R_C \tag{7.14}$$

where :

ρ : specific resistivity
L : drawn resistor length
ΔL : total effective length reduction from end effects
W : drawn resistor width
ΔW : total encroachment
R_C : ohmic contact resistance on each side

In order to make this treatment as general as possible, we chose to separate the end effects (ΔL) from the actual contact resistance, R_C. Depending upon the architecture of the resistor and particularly the contact structure, either component might be the dominant one; the ohmic contact resistance or the current crowding end effects. The process might have metal contact, poly contacts, or might be utilizing tungsten plugs, and whatever technology is used will determine which component would be dominant. For some structures where the encroachment is negligible it is impossible to separate the two effects, and an overall effective contact resistance will be extracted.

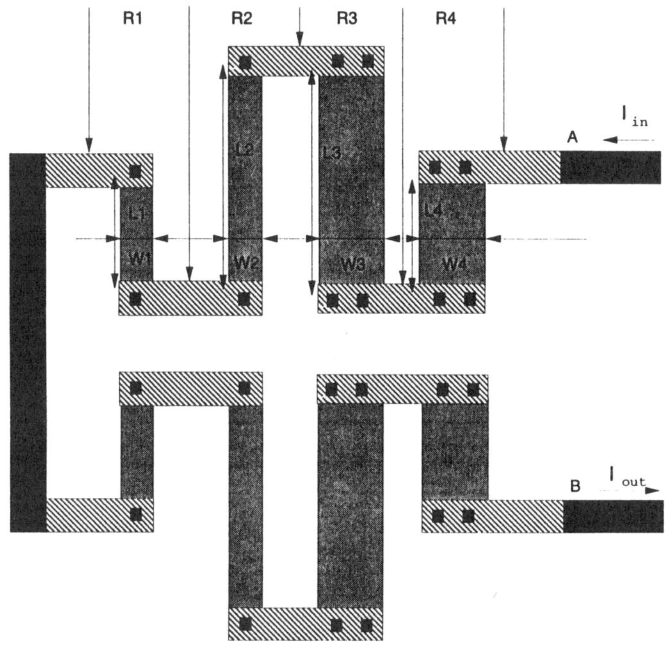

Figure 7.11: Resistance Extraction Structure

The test structure consists of two identical sections, each containing four resistors in series as shown in Figure 7.11. Each section by itself yields enough information to extract all parameters given by Equaion (7.14); the second section is included for matching and measurement reliability purposes.

This structure was designed for on-chip parameter extraction using a probe card and a fully automated probing facility using twelve probes per measurement. Two probes are used for forcing the biasing current, and ten probes are used as simple voltage sensors, thus eliminating errors introduced by the probe contact resistances. In order to enable us to extract all parameters given in Equation 7.14, we need to layout two resistors with the same length, and two with the same width. This will isolate the termination effects, ΔL, and simplify the calculations. The resistors of identical width will also have identical contacts, and if we pick all resistors sufficiently large, we can assume the encroachment to be the same for all resistors. This gives the following simplifications:

$$
\begin{aligned}
W_1 = W_2 = W_A & \qquad W_3 = W_4 = W_B \\
R_{C1} = R_{C2} = R_{CA} & \qquad R_{C3} = R_{C4} = R_{CB} \\
\Delta L_1 = \Delta L_2 = \Delta L_A & \qquad \Delta L_3 = \Delta L_4 = \Delta L_B \\
L_1 = L_4 = L_A & \qquad L_2 = L_3 = L_B
\end{aligned}
\tag{7.15}
$$

222

$$\Delta W_1 = \Delta W_2 \quad = \quad \Delta W_3 = \Delta W_4 = \Delta W$$
$$\rho_1 = \rho_2 \quad = \quad \rho_3 = \rho_4 = \rho$$

Thus we get the following expressions for the four resistances:

$$R_1 \quad = \quad \rho\frac{L_A + \Delta L_A}{W_A + \Delta W} + 2R_{CA} \qquad (7.16)$$
$$R_2 \quad = \quad \rho\frac{L_B + \Delta L_A}{W_A + \Delta W} + 2R_{CA}$$
$$R_3 \quad = \quad \rho\frac{L_B + \Delta L_B}{W_B + \Delta W} + 2R_{CB}$$
$$R_4 \quad = \quad \rho\frac{L_A + \Delta L_B}{W_B + \Delta W} + 2R_{CB}$$

From these equations we can find the resistivity, ρ, and the encroachment, ΔW, directly, given by the following equations:

$$\rho \quad = \quad \frac{R_4 - R_3}{L_A - L_B}(W_B + \Delta W) \qquad (7.17)$$

$$\Delta W = \frac{(R_2 - R_1)W_A + (R_4 - R_3)W_B}{(R_3 - R_4) + (R_1 - R_2)} \qquad (7.18)$$

In order to make these expressions well behaved, the physical dimensions should be chosen as to ensure that the following inequalities hold:

$$R_2 \neq R_1 \quad R_4 \neq R_3 \quad L_B \neq L_A \quad W_B \neq W_A \qquad (7.19)$$

Comparing the restrictions introduced by Equation 7.19, to the choices picked in Equation 7.15, it is clear that all the restrictions are satisfied. It might not be immediately obvious as to why the widths of the two sets have to be different, but looking at Equation 7.16, we find that if W_A is equal to W_B, we would not be able to separate ρ and ΔW. For good resolution and good measurement accuracy it would be advantageous to pick the dimensions as follows:

$$L_B = 2L_A \qquad W_B = 2W_A \qquad (7.20)$$

This yields the following approximate resistance ratios based on the drawn dimensions only:

$$R_2 = 2R_1 = 2R_3 = 4R_4 \qquad (7.21)$$

Thus all the restriction given by Equation 7.19 is satisfied with good margins.

In order to extract the contact resistance and the end effects, two additional equations are required. If the two resistor pairs are laid out such that each resistor having twice the width is simply implemented by two of the thinner ones in parallel, the resistance associated with the contact should be cut in half. The termination effects on the effective length should be identical for the two pairs. Thus we get the required additional two equations:

$$W_B = 2W_A \quad \Rightarrow \quad \Delta L_A \approx \Delta L_B = \Delta L \tag{7.22}$$

$$R_{CA} \approx 2R_{CB} = R_C$$

Solving Equation 7.16 with these additional constraints, the contact resistance and effective length shortening are given by the following equations:

$$\Delta L \;=\; \frac{(R_1 - 2R_3)(W_A + \Delta W)(W_B + \Delta W)}{\rho(W_B - 2W_A - \Delta W)} \tag{7.23}$$

$$- \frac{\rho[2L_B(W_A + \Delta W) - L_A(W_B + \Delta W)]}{\rho(W_B - 2W_A - \Delta W)}$$

$$R_C = R_2 - \rho\frac{L_B + \Delta L}{W_A + \Delta W} \tag{7.24}$$

For resistor types in which the lateral encroachment is negligible, ($\Delta W \approx 0$), Equations 7.23 and 7.24 do not provide accurate values for ΔL and R_C; instead, these effects must be combined into an effective contact resistance, R_{Ceff}. In this case we obtain the following expression for the overall effective contact resistance:

$$R_{CBeff} = \rho\frac{\Delta L_B}{W_B} + 2R_{CB} = 2R_4 - R_3 \tag{7.25}$$

$$R_{CAeff} = \rho\frac{\Delta L_A}{W_A} + 2R_{CA} = 2R_1 - R_2 \tag{7.26}$$

This measurement procedure is carried out on both sets of resistors, giving separate values of ρ, ΔL, ΔW, and R_C for each section. These separate values are subtracted to give the parameter mismatches between the two sections ($\delta\rho = \rho_{sectA} - \rho_{sectB}$, $\delta L = \Delta L_{sectA} - \Delta L_{sectB}$ etc.). If this process is repeated over a statistically large number of resistor pairs, the resulting estimates for $\sigma(\delta\rho)$, $\sigma(\delta L)$, $\sigma(\delta W)$, and $\sigma(\delta R_C)$ will provide an indication of which parameter most severely limits resistance matching.

7.5.2 Resistance Mismatch Modeling

In order to obtain an expression suitable for a statistical treatment of resistor mismatch using the model given in Equation 7.14, we will obtain an expression for the mismatch between a pair of similar resistors, R_1 and R_2, in terms of the small random mismatches in their parameters, $\delta\rho$, δL, δW, and δR_C. From Equation 7.14, R_1 and R_2 are given by the following expressions:

$$R_1 \;=\; (\rho + \frac{\delta\rho}{2})\frac{L + \delta L/2}{W + \delta W/2} + 2R_C + \delta R_C$$

$$R_2 \;=\; (\rho - \frac{\delta\rho}{2})\frac{L - \delta L/2}{W - \delta W/2} + 2R_C - \delta R_C \tag{7.27}$$

224

Neglecting higher order terms and assuming $\delta W \ll W$ the expression for R_1 may be simplified as follows:

$$R_1 = (\rho + \delta\rho/2)\frac{L + \delta L/2}{W + \delta W/2} + 2R_C + \delta R_C \qquad (7.28)$$

$$= \frac{(\rho + \delta\rho/2)(L + \delta L/2)}{W}\frac{1}{1 + \frac{\delta W}{2W}} + 2R_C + \delta R_C$$

$$\approx \frac{(\rho + \delta\rho/2)(L + \delta L/2)}{W}(1 - \frac{\delta W}{2W}) + 2R_C + \delta R_C$$

$$\approx (\rho + \delta\rho/2)(\frac{L}{W} + \frac{\delta L}{2W} - \frac{L\delta W}{2W^2}) + 2R_C + \delta R_C$$

with a similar expression for R_2. Using these expressions and neglecting higher order terms we obtain the following expressions for the absolute and relative mismatch:

$$\delta R = \rho\frac{\delta L}{W} - \rho\frac{L\delta W}{W^2} + \delta\rho\frac{L}{W} + 2\delta R_C \qquad (7.29)$$

$$\frac{\delta R}{R} = \rho\frac{\delta L}{RW} - \rho\frac{L\delta W}{RW^2} + \delta\rho\frac{L}{RW} + 2\frac{\delta R_C}{R} \qquad (7.30)$$

If we assume $\rho L/W \gg 2R_C$, or $R = \rho L/W$, this expression simplifies to

$$\frac{\delta R}{R} = \frac{\delta L}{L} - \frac{\delta W}{W} + \frac{\delta\rho}{\rho} + 2\frac{\delta R_C}{R} \qquad (7.31)$$

Clearly, for optimum matching, we need to use resistors with large resistance and large physical dimensions.

If a large database of mismatch measurements is available, the probability density functions for the parameter mismatches can be determined, a statistical model can be established, and optimization and worst case analysis performed using this model [4], [8].

A test structure has been discussed that enables the circuit designer and process engineer to determine how closely two identical resistors can be matched or a given ratio be achieved, and that also gives insight into where the mismatch is introduced. This test structure and extraction methodology can be applied to both diffusion and ploysilicon resistor.

The structure is designed to allow the extraction of the sheet resistance, the lateral encroachment, and the effective contact resistance. Further, if the encroachment is non-negligible, the contact resistance can be separated into separate components due to current crowding and actual ohmic resistance.

7.6 MOS Capacitance Characterization Technique

The measurement of MOS transistor capacitance is of particular interest for analog circuit design [16]. The information can be used to complement the I-V parameters and thus to provide more accurate circuit simulations. Despite the importance of intrinsic capcitances information, it is not widely studied. This is mainly due to two reasons [15]. First, the existing capacitance meter is not designed for more than two terminal measurement. Therefore, the measurement for a MOS transistor, which is a four terminal device, is not trivial. Furthermore, the requirement for the resolution of the measuring meter is high; that is, one femtofarad or better. These reasons make the measurement of transistor intrinsic capacitances difficult. One way to handle this obstacle is to use "on chip" methods. A special test circuit, including a reference capacitor, is created in the proximity of the devices that are tested. The voltage across the reference capacitor is measured and the capacitances of the transistor are then extracted.

The "on chip" method improves the resolution but the methodolgy requires extra test circuits [17]. Also it relies on both of the test circuits and the device under test to be functional. In addition, it may be impractical to incorporate the test circuit beside a real design. Furthermore, extensive calculation is needed in the "on chip" measurement methodology. The following section presents a direct capacitance measurement technique [17]. A direct measurement technique is more desirable because no extra circuits are needed. In addition, no extensive calculation is required, thus the result can be obtained more efficiently.

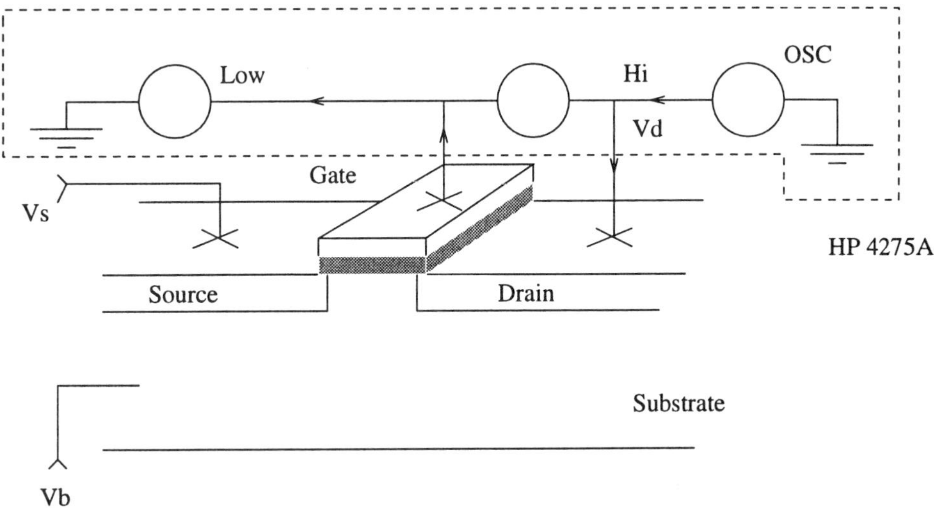

Figure 7.12: Capacitance Measurement Configuration

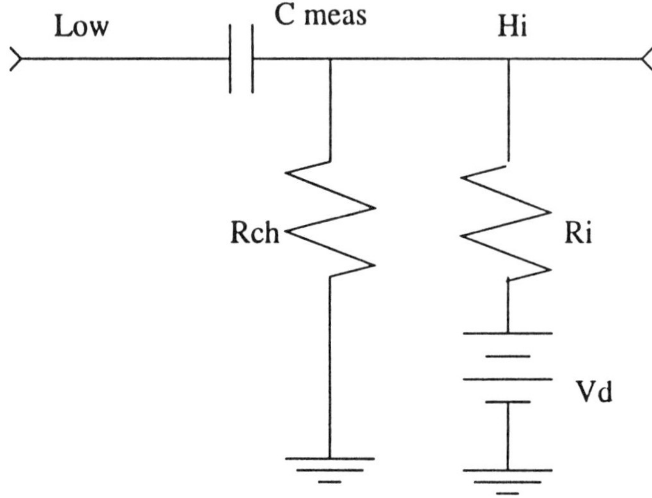

Figure 7.13: Channel Resistance Effect on Drain Bias

7.6.1 Direct Intrinsic Capacitance Measurement

The resolution of HP 4275A can be as high as hundredth of a femtofarad. The meter is regarded through two vector meters (i.e. voltmeter and ammeter) from which the capacitance between the "HI" and "LOW" terminals can be extracted. The configruation that is used to extract the intrisinic capacitances of a MOS transistor is shown in Figure 7.12. Additional voltage sources are used to bias the transistor into different operation regions. Since there is one dc source at the "HI" terminal, two independent dc sources are connected to the source terminal and the substrate node of the device.

A major problem in using the HP 4275A to extract the capacitances of a MOS transistor is the channel resistance effect on the actual drain voltage. The equivalent channel resistance is neither in series nor in parallel to the capacitance under measurement. This can be further explained using Figure 7.13. R_{ch} is the equivalent channel resistance and R_i is the output resistance of the HP 4275A. Note that R_i is large compared to R_{ch}. Since the value of R_{ch} will vary with the number of minority carriers in the channel, any changes in R_{ch} will modify the drain bias. To overcome this problem, the drain voltage is adjusted continuously so as to provide the correct bias.

The direct measurement of MOS capcitance consists of three components. They are the parasitic capacitance, the overlap and outer fringing capacitance and the intrisinc capacitance. The parasitic capacitance includes the capacitance such as from bonding pad, and metallic via. The overlap capacitance comes from the encroachment of drain/source region in the channel region while the outer fringing capacitance is from the fringing field. These two components are assumed to be independent of biasing conditions.

The intrinsic capacitances of a MOS device can be separated into two components. The first component is the inner fringing capactiance. It exists when the gate oxide is not covered by mobile charges completely at the channel surface. This component is dominant only when the device is in saturation. It is shown that the effect of the inner fringing capacitance decreases in the subthreshold and accumulation region. Therefore, the intrinsic capacitance of a device can be extracted by the difference between the measured capacitance and the data measured when the device is in accumulation. Based on this method, the gate and substrate instrinsics can be extracted.

Bibliography

[1] P. E. Allen and D. R. Holberg, "CMOS Analog Circuit Design", Holt, Rinehart and Winston Inc. 1987.

[2] P. R. Gray and R. G. Meyer, "Analysis and Design of Analog Integrated Circuits", John Wiley and Sons, 1984.

[3] G. Massobrio and P. Antognetti, "Semiconductor Device Modeling with SPICE," 2nd Edition, McGraw Hill, 1993.

[4] C. Abel, C. Michael, M. Ismail, C. Teng, and R. Lahri, "Characterization of transistor mismatch for statistical CAD of submicron CMOS analog circuits," in Proceedings of the Internationa; Symposium on Circuits and Systems, pp. 1401 - 1404, 1993.

[5] F. Larsen, "Bipolar Device Characterization and Design in CMOS Technologies for the Design of High-Performance Low-Cost BiCMOS Analog Integrated Circuits", PhD Dissertation, The Ohio State Unversity, 1994.

[6] I. E. Getreu, "Modeling the bipolar transistor", Textronics Inc., Beaverton, Oregon, 1976.

[7] C. Michael and M. Ismail, "Statistical Modeling for CAD of VLSI Circuits," Kluwer Academic Publisher, Boston, 1993.

[8] C. Michael and M. Ismail, "Statistical Modeling of Device Mismatch for Analog MOS Integrated Circuits," IEEE J. Solid-State Circuits, vol. 27, pp.154-166, Feb 1992.

[9] M. Pelgrom, A. Duinmaijer, and A. Welbers, "Matching Properties of MOS Transistors", *IEEE Journal of Solid State Circuits*, pp. 1433 - 1440, October 1989.

[10] H.Y. To, "Statistical Analysis and Design Techniques for Analog VLSI Circuits", PhD Dissertation, The Ohio State Unversity, 1995.

[11] F. Larsen, A. Iranmanesh, and M. Ismail, "A versatile test structure for measuring resistance matching properties." Accepted for publication in the IEEE Jouranl of Solid-State Circuits.

[12] S. Zarabadi, M. Ismail, and F. Larsen, "Basic BiCMOS Circuit Techniques", Ch. 5 in "Analog VLSI: Signal and Information Processing", edited by M. Ismail and T. Fiez, McGraw Hill, New York 1994.

[13] L. J. van der Pauw, " A Method of Measuring Specific Resistivity and Hall Effects of Discs of Arbitrary Shape", *Phil. Res. Rep. 13*, pp. 1-9, February 1958.

[14] S.J. Proctor, L.W. Linholm, and J.A. Mazer, " Direct Measurement of Interfacial Contact Resistance, End Resistance, and Interfacial Contact Layer Uniformity ", *IEEE Trans. Electron Devices*, ED-30, pp. 1535 - 1542, November 1983.

[15] W. Lin, and P. Chan, " On the Measuremnet of Parasitic Capacitances of Device with More than Two External Terminals Using an LCR Meter", *IEEE Trans. Electron Devices*, ED-38, pp. 2573 - 2575, November 1991.

[16] D. Flandre, F. van de Wiele, P. G. A. Jespers, and M. haond, " Measurement of Intrinsic Gate Capacitances of SOI MOSFET's", *IEEE Trans. Electron Devices Letters*, EDL-11, pp. 291 - 293, January 1990.

[17] K.C.-K. Weng, and P. Tang, "A Direct Measurement Technique for Small Geometry MOS Transistor Capcitances", *IEEE Trans. Electron Devices Letters*, EDL-6, pp. 40 - 42, January 1985.

Index